The U.S. Coal Industry

The MIT Press Energy Laboratory Series

1. Chemical Equilibria in Carbon-Hydrogen-Oxygen Systems
Robert E. Baron, James H. Porter, and Ogden H. Hammond, Jr., 1976

2. Electric Power in the United States: Models and Policy Analysis
Martin L. Baughman, Paul L. Joskow, and Dilip P. Kamat, 1979

3. The U.S. Coal Industry: The Economics of Policy Choice
Martin B. Zimmerman, 1981

The U.S. Coal Industry:
The Economics of Policy Choice

Martin B. Zimmerman

The MIT Press
Cambridge, Massachusetts
London, England

The book was set in English Times by A&B Typesetters, Concord, N.H. and printed and bound by Halliday Lithograph in the United States of America.

Library of Congress Cataloging in Publication Data

Zimmerman, Martin B.
  The U.S. coal industry

  (The MIT Press energy laboratory series ; 3)
  Includes bibilographies and index.
  1. Coal trade—United States. 2. Coal trade—United States—Mathematical models. I. Title. II. Series: MIT Press energy laboratory series ; 3.
HD9545.Z55   338.2 '724 '0973   81-1124
ISBN 0-262-24023-0   AACR2

To Nancy

Contents

List of Tables   x

Acknowledgments   xv

1

Introduction

1.1 The Coal Environment   3

1.2 The State of Regulation   11

1.3 Nuclear Power and Coal   13

1.4 The Scramble for Rents   14

1.5 The Method of Analysis   15

References   16

2

The Models

2.1 Model Overview   17

2.2 The Coal Supply Model   24

2.3 The Transport Model   42

2.4 The Coal Demand Model   45

2.5 The Links between the Supply and Demand Models   50

References   56

3

Basic Forces

3.1 Economic Factors in the Industry   61

3.2 The Effects of Depletion and Demand Shifts   64

3.3 Relative Factor Prices   72

3.4 Transportation Cost Increases   76

3.5 Supply Developments   79

3.6 The Impact of Factor Price Changes on Demand   82

3.7 User Costs   91

3.8 Basic Forces and Demand Policy Response   93

References   98

4

The Economics of Environmental Trade-Offs

4.1 Sulfur Emission Standards and the Price of Coal   101

4.2 The Effects of Alternative Sulfur Emission Levels on Fuel Choice   106

4.3 The Costs of Sulfur Reduction   116

4.4 Sulfur Regulations and Oil Imports   128

4.5 Reducing Western Mining   129

4.6 Environmental Trade-Offs and Coal Policy   147

References   150

5

The Coal Industry without Nuclear Energy

5.1 The Electric Utility Model for Predicting Nuclear Power Generation   152

5.2 The Nuclear Moratorium   155

5.3 Interactions of Coal and Nuclear Policy   161

5.4 The Trade-Offs   168

References   169

6

The Future of Coal and the Alternatives

6.1 Economic Forces Influencing the Industry   170

6.2 Coal Policy and Oil Imports   173

6.3 The Three Safety Valves and the Question of Timing   175

References   178

Contents

Appendixes

A Distribution of Strippable Reserves in Appalachia   179

B Base-Case Assumptions in the Coal Model   182

C Strip-Mine Reclamation Costs in the Coal Model   189

D 1975 Transportation Costs   190

Notes   192

Index   203

List of Tables

1.1 Coal consumption in the United States and exports, 1946 to 1974   6

1.2 Production of coal by state   8

1.3 Percentage of output accounted for by deep mining for selected eastern states, 1950 to 1978   9

2.1 Definition of coal regions   21

2.2 Sulfur categories used in model   21

2.3 Underground-mining expenditures as a function of sections   32

2.4 Strip-mine expenditures   36

3.1 Estimated percent increases in cost over time at current rates of output for eastern deep coal   62

3.2 Estimated percent increases in cost over time for midwestern and western surface-mined coal at current rates of output   62

3.3 Trends in labor productivity and labor costs   63

3.4 Regional production under the assumptions of constant factor prices and no-sulfur contraints   65

3.5 Distribution pattern of tons of coal shipped from Montana-Wyoming to demand regions, selected years   66

3.6 Consumption of midwestern coal under assumptions of constant factor prices and no sulfur regulations   66

3.7 Shipments to the south Atlantic demand region by supplying region   67

3.8 Regional production under the assumptions of constant factor prices and no-sulfur constraints, steam market only   68

3.9 Delivered cost of highest-sulfur coal to south Atlantic markets   69

3.10 Delivered prices of coal ($/million Btu) with constant factor prices, no sulfur regulations   70

3.11 Regional production with constant factor prices and 1.2 lb/million Btu sulfur standard in 1985   71

3.12 Delivered cost of coal with constant factor prices, 1.2 lb SO$_2$ sulfur constraint   73

3.13 Fraction of total cost accounted for by labor, materials, and supplies   74

3.14 Recent behavior of the mining machinery price index   75

3.15 Regional production under assumptions of changing relative prices and NSPS   76

3.16 Recent behavior of rail cost indexes   77

3.17 Regional coal production with transport cost increases of 3 percent per year to 1985   77

3.18 Comparison of coal distribution patterns in 1990 with and without transport cost increases   78

3.19 Comparison of coal distribution patterns in 2000 with and without transport cost increases   80

3.20 Average annual growth rate of electricity demand   83

3.21 Generation capacity   84

3.22 Delivered coal prices for selected regions, 1980 to 2000   86

3.23 Average national electricity prices under alternative scenarios   88

3.24 Industrial coal demand   90

3.25 Industrial oil and gas consumption   90

3.26 Mine-mouth cost of coal ($/ton in nominal $) cumulative production, real cost increase, royalty for selected regions, without sulfur regulations   94

3.27 Mine-mouth cost of coal for selected regions with sulfur regulations of 1.2 lb SO$_2$ per million   96

4.1 Delivered price of low-sulfur coal   102

4.2 Delivered price of high-sulfur coal   104

4.3 Industrial use of coal   106

4.4 Electrical generation capacity by type of plant   108

4.5 Scrubbing capacity by region   109

4.6 Total demand for electricity   109

4.7 Regional coal production under alternative sulfur emission levels   110

4.8 Total shipments of coal   112

4.9 Price of electricity   118

4.10 Demand for electricity   120

4.11 National electricity demand   120

4.12 Cost of alternative sulfur standards to consumers of electricity   122

4.13 Cost to electric consumers of alternative sulfur levels by region in 1990   124

4.14 Loss to industrial consumers of a 1.2 lb SO_2 sulfur standard in 1980   124

4.15 Change in rents to coal producers by region and sulfur content in 1990 due to 1.2 lb SO_2 standard   126

4.16 Tax revenues   126

4.17 Industrial use of gas, oil, and electricity   128

4.18 Industrial demand for energy   129

4.19 Production by region under 1.2 lb SO_2 sulfur standard with BACT   132

4.20 Coal distribution in 1995 with BACT and 1.2 lb SO_2 sulfur standard   134

4.21 Coal distribution in 2000 with BACT and 1.2 lb SO_2 sulfur standard   134

4.22 The effect of BACT on the delivered cost of low-sulfur coal   136

4.23 Electricity prices   136

4.24 Cumulative output, 1975 to 2000   138

4.25 Federal ownership of western lands   138

4.26 Production by region under a leasing moratorium   140

4.27 Delivered price of coal under a leasing moratorium   140

4.28 Utility coal consumption by sulfur content under a coal-leasing moratorium   142

4.29 Electricity prices under reference case and leasing moratorium   142

4.30 Industrial coal use under a leasing moratorium   144

4.31 Regional output under western transport rate increase   147

4.32 Distribution of coal with high western transport rates in 2000   148

4.33 Regional delivered coal prices under higher transport costs   148

5.1 Comparison of the cost of generating base-load electricity by coal and nuclear power plants   153

5.2 Sensitivity of generation capacity to reduced capacity factors and higher capital costs   154

5.3 Coal production by region under a nuclear moratorium   155

5.4 Coal distribution under a nuclear moratorium   158

5.5 Cumulative coal output, 1975 to 2000, with and without a nuclear moratorium   158

5.6 Mine-mouth prices under a nuclear moratorium compared to reference case in 2000   160

5.7 Electricity prices under alternative scenarios   162

5.8 Regional coal production under a nuclear moratorium without BACT   165

5.9 Regional output under a nuclear moratorium without sulfur standards and BACT   166

5.10 Regional output under a nuclear moratorium and a western leasing moratorium   167

5.11 The demand for electricity with alternative coal policies under a full nuclear moratorium   167

5.12 Coal consumption under various coal policies and a nuclear moratorium   167

6.1 Cumulative output by region, 1975 to 2000   171

6.2 U.S. industrial coal consumption in selected scenarios compared to Department of Energy forecasts   172

6.3 The estimated cost of synthetic fuels   172

6.4 Forecasts of steam coal exports   174

6.5 Oil consumption under selected scenarios   174

6.6 Generating capacity and coal consumption under alternative assumptions in 2000   177

B.1 Tax rates   182

B.2 Assumed real rates of escalation   182

B.3 SO_2 constraints on incremental coal   183

B.4 Btu per ton of coal   183

B.5 Exogenous levels of metallurgical and export coal   183

B.6 Reference case assumptions of the electric utility model: expected capital costs for plants   184

B.7 Reference case assumptions of the electric utility model: realized capital cost   184

B.8 Reference case assumptions of the electric utility model: regional capital cost multipliers   184

B.9 Reference case assumptions of the electric utility model: heat rates   185

B.10 Reference case assumptions of the electic utility model: operation and maintenance costs   185

B.11 Reference case assumptions of the electric utility model: maximum capacity factor achievable   185

B.12 Reference case assumptions of the electric utility model: other fuel prices   186

B.13 Reference case assumptions of the electric utility model: fuel transport charges   186

B.14 Reference case assumptions of the electric utility model: financial parameters   187

B.15 Reference case assumptions of the electric utility model: nuclear capacity constraints   187

C.1 Cost of reclamation   189

D.1 Transportation costs, including rents, in 1975   190

Acknowledgments

This book is the product of several years of research activity. Consequently I have accumulated a number of debts along the way. Many people have provided critiques and guidance. I owe a special debt of gratitude to M. A. Adelman and Richard L. Gordon, who provided valuable criticisms throughout the development of the ideas incorporated in this book. Jerry Hausman, Martin Baughman, and Dilip Kamat also helped in various ways. In the process of developing the coal model I have had excellent research assistants. I wish to thank Michael Baumann, in particular, who helped at every stage of the research and authored an appendix in this book. He and George Rozanski were instrumental in the integration of the models. Also contributing to the overall effort were Christopher Alt and Alec Sargent.

The work was supported by grants from the Council on Wage and Price Stability, the Department of Energy, and, most important, by the Center for Energy Policy Research at the Massachusetts Institute of Technology. I am grateful to all these institutions.

Peter Heron skillfully edited the manuscript, and Alice Sanderson worked wonders with the word processor.

Finally, my greatest debt is to Nancy Reiner Zimmerman.

The U.S. Coal Industry

1   Introduction

The recent dramatic rise in the price of oil has increased expectations for the U.S. coal industry. Policy planners hope coal will supplant oil in industrial and utility use, thereby reducing the dependence on imported oil. The Carter administration called for a doubling of coal output by 1985 and the creation of a coal-based synthetic fuel industry by 1990. Coal is seen as a backstop for the faltering nuclear power industry.

The resurgence of the coal industry actually antedates the OPEC oil embargo that took place during the end of 1973 and at the beginning of 1974. The post-World War II decline of the industry had ended by the early 1960s. Electricity production was increasing rapidly, and the growth in coal consumption by electric utilities offset the loss of markets in the household and transportation sectors. The demand by utilities received a boost from natural gas price controls. The controls caused shortages that by the late 1960s eliminated gas as a source of fuel for large new installations. Optimism about nuclear power also was waning in the early 1970s, maintaining the promise of future coal expansion; see Zimmerman and Ellis [7].[1] The oil price increases of 1973 and 1974, along with import disruptions, led policy makers to conclude that coal would and should play an even greater role in meeting national energy needs.

The increase in actual production along with optimism about the future of coal was occurring at the same time that environmental policies were becoming more stringent. The events of 1973 and 1974 came at a time of tight environmental standards. The Clean Air Act of 1967 and its amendments of 1970 placed restrictions on emissions of sulfur dioxides, regulations which were scheduled to become tighter over time. In 1969 the Coal Mine Health and Safety Act was passed; by 1973 the coal industry was still adjusting to those regulations. Finally, Congress was threatening to regulate strip-mining, and many states had already begun to pass their own reclamation legislation.

Against this background of environmental regulation, it is not surprising that there was appreciable doubt as to whether the industry could achieve the levels of growth envisioned. These doubts were not assuaged by developments that took place throughout the 1970s and that continue to this day. Environmental regulation has become even more stringent, further increasing doubt over the earlier optimism about coal use; see Gordon [4]. At present pressures are building to relax environmental standards in order to increase the use of coal.

The issue is partly one of cost. Those who see little difficulty ahead feel that the cost of complying with environmental regulations is not large enough to affect the choice between coal and other fuels. Furthermore, they argue, this is an efficient policy—the price of coal should reflect damage to the environment. A second group emphasizes the cost of complying with environmental regulation. It claims that the federal government is pursuing contradictory policies. On the one hand, laws are passed calling for conversion to coal, and, on the other, environmental standards are established that make the use of coal extremely expensive. The latter group presents a choice: either lower environmental standards, or accept the fact that coal will play a more limited role in satisfying national energy needs. This group emphasizes the social cost of imported oil.

In an effort to limit oil imports, the government has begun to assume a role in directing fuel choice by private firms. Legislation prohibits the use of oil in new major burning installations. The use of gas is also restricted; see Department of Energy [6].[2] Public reaction and regulatory delay have created fears that, once built, nuclear plants might not be operated. This has effectively eliminated nuclear power from competition with coal; see Zimmerman and Ellis [7]. Recent developments at the Three Mile Island station of General Public Utilities have confirmed this trend. While all these developments favor the use of coal, they make the environmental trade-offs with respect to coal more acute. The cost of correcting environmental damage from coal mining depends importantly upon the role of nuclear power.

In the midst of the debate over imports and the environment, income distribution goals exert a powerful influence on policy. Appalachia has seen its traditional markets adversely affected by environmental regulations. That segment of the industry is calling for legislation to arrest the decline of its markets. In sum the evolution of the coal industry will be affected both by relative costs and policies responding to different and often competing interest groups.

The processes that determine policy are not centralized. Rather, decisions are made by various agencies of the federal government, each of which responds to different constituencies. Local governments, through taxing power, can and will have an important impact on coal development. The policies that actually emerge will do so in a decentralized manner. Local governments not willing to accept the implications of national policy will react with countervailing local policies. Federal agencies will act independently of each other. Trade-offs will be made in a disorganized way with little consideration of how various decisions interact.

This book will attempt to take the larger view and examine the problem from a national perspective. How do various policy initiatives interact? What are the terms of the trade-offs? Must we sacrifice environmental goals to increase coal use? What will it cost to achieve a clean environment? How will a nuclear moratorium affect the coal industry?

1.1 The Coal Environment

The coal industry can be divided broadly into three major producing areas: Appalachia, the midwest or Illinois Basin, and the states west of the Mississippi River. Figure 1.1 presents a map of the coal fields of the continental United States.

Two key aspects of coal production bear significantly on policy: mining technology and coal quality. In the Appalachian fields surface mining accounts for approximately 40 percent of total production. In the Illinois Basin surface mining accounts for 52 percent of total production, although current trends are for increased deep mining. The west is primarily a strip-mining area; the only significant amount of deep mining takes place in Colorado, Utah, and the southwestern portions of Wyoming.

The most important quality dimensions of coal are its heating value and sulfur content. In general the heating value per ton of coal declines from east to west. For example, the average heating value of a ton of coal in Montana and Wyoming is about 30 percent less than in the Appalachian coal fields. As to sulfur content, midwest coal is generally high (above 2 percent) and western coal generally low (below 1 percent). The sulfur content of Appalachian coal is quite variable from seam to seam.

Since the main policy debates revolve around the regulation of strip mining and the control of sulfur emissions, these basic geographical and geological facts set the stage for policy determination. Policies aimed at either of these factors have important implications for regional coal production and use.

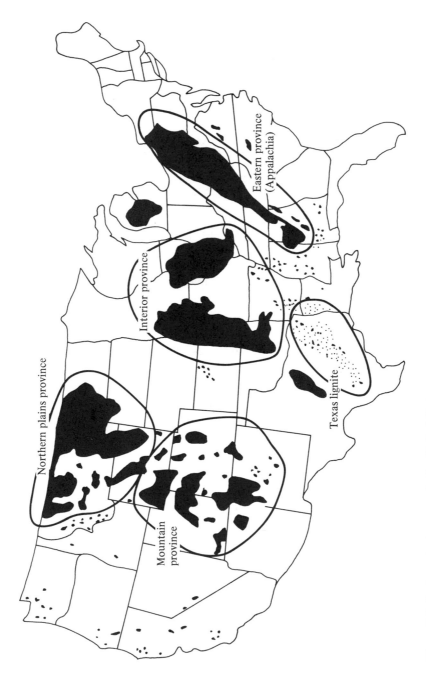

Figure 1.1 Coal-producing areas in the United States

Consumption Patterns

After World War II the coal industry lost two important markets to oil. The railroads converted from coal to diesel fuel, and the residential and commercial markets converted to fuel oil. Total U.S. consumption of coal declined from 546 million tons in 1947 to 347 million tons per year in 1961 (see table 1.1). The railroads accounted for a loss in consumption of over 100 million tons per year. Consumption by railroads declined from 109 million tons in 1947 to 2 million tons in 1960. Retail deliveries declined from 97 million tons in 1947 to 30 million tons in 1960. Industrial use (other than electric utilities) declined by about 40 million tons per year. The only sector showing significant growth in coal consumption was the electric utility sector, in which consumption increased from 86 million tons to 180 million tons in 1961.

These trends in consumption patterns continued throughout the 1960s and 1970s. Total consumption began to rise in the 1960s, and by 1973 total consumption had again reached the 1947 level. The engine of growth was again the electric utility sector, whose level of consumption grew from 180 million to 319 million tons over the 1960s and reached 476 million tons in 1978.[3] Forecasters see the impetus for future growth in the coal industry coming from continued expansion of coal burning in the electric utility sector.

Production Trends

In response to sulfur pollution regulations, changing factor prices, and demand patterns, coal production in the United States has been slowly shifting toward the west (see table 1.2). In 1960 the producing states east of the Mississippi River accounted for 95 percent of total production. By 1970 this had declined to 85 percent, and in 1978 eastern states accounted for 77 percent of total production. This trend shows every indication of continuing.

Part of this movement to the west has been caused by the relative attractiveness of strip mining. Technological developments, factor price changes, and regulatory developments have over time lowered the relative cost of strip mining. The development of large draglines and other earthmoving equipment made large-scale stripping possible and relatively cheap. Rising real wages and declining labor productivity since 1970 also encouraged capital-intensive strip mining over labor-intensive deep mining. In many eastern areas new large strip reserves were unavailable so the large reserves of the west became more attractive.

Table 1.1 Coal consumption in the United States and exports 1946 to 1974 (thousand tons)

Year	Electric power utilities	Railroads (class I)	Coking coal	Steel and rolling mills
1946	68,743	110,166	83,288	12,151
1947	86,009	109,296	104,800	14,195
1948	95,620	94,838	107,306	14,193
1949	80,610	68,123	91,236	10,529
1950	88,262	60,969	103,845	10,877
1951	101,898	54,005	113,448	11,260
1952	103,309	37,962	97,614	9,632
1953	112,283	27,735	112,874	8,764
1954	115,235	17,370	85,391	6,983
1955	140,550	15,473	107,377	7,353
1956	154,983	12,308	105,913	7,189
1957	157,398	8,401	108,020	6,938
1958	152,928	3,725	76,580	7,268
1959	165,788	2,600	79,181	6,674
1960	173,882	2,101	81,015	7,378
1961	179,629	b	73,881	7,49
1962	190,833	b	74,262	7,319
1963	209,038	b	77,633	7,401
1964	223,032	b	88,757	7,394
1965	242,729	b	94,779	7,466
1966	264,202	b	95,892	7,117
1967	271,784	b	92,272	6,330
1968	294,739	b	90,765	5,657
1969	308,462	b	92,901	5,560
1970	318,921	b	96,009	5,410
1971	326,280	b	82,809	5,560
1972	348,612	b	87,272	4,850
1973	386,879	b	93,634	6,356
1974	390,068	b	89,747	6,155

Source: USBM in *Minerals Yearbook* and *Coal-Bituminous and Lignite*.

[a] Includes bunker fuel.

[b] Data included in other manufacturing and mining industries.

[c] Beginning in 1970, the data listed in these two columns were combined.

Cement mills	Other manufacturing and mining industries[a]	Retail deliveries to other consumers	Total U.S. consumption	Exports
6,990	120,364	98,684	500,386	41,197
7,919	127,015	96,657	545,891	68,667
8,546	112,612	86,794	519,909	45,930
7,966	98,685	88,389	445,538	27,842
7,923	97,904	84,422	454,202	25,468
8,507	105,408	74,378	468,904	56,722
7,903	95,476	66,861	418,757	47,643
8,167	96,999	59,976	426,798	33,760
7,924	78,359	51,798	363,060	31,041
8,529	91,110	53,020	423,412	51,277
9,026	94,772	48,667	432,858	68,553
8,633	88,566	35,712	413,668	76,446
8,256	82,327	35,619	366,703	50,291
8,510	74,365	29,138	366,256	37,253
8,216	77,432	30,405	380,429	36,541
7,615	78,050	27,735	374,405	34,970
7,719	79,453	28,188	387,774	38,413
8,138	83,467	23,548	409,225	47,078
8,679	83,639	19,615	431,116	47,969
8,873	86,269	19,048	459,164	50,181
9,149	89,941	19,965	486,266	49,302
8,922	84,009	17,099	480,416	49,528
9,391	83,054	15,224	498,830	50,637
9,131	78,557	14,666	507,275	56,234
	83,207[c]	12,072	515,619	70,944
	68,862[c]	11,351	494,862	56,633
	67,294[c]	8,748	516,776	55,997
	60,953[c]	8,200	556,022	52,870
	57,899[c]	8,840	552,709	59,926

Table 1.2 Production of coal by state (thousand tons)

	1960	1970	1975	1977
Alabama	12,600	20,560	22,644	21,220
Kentucky	64,000	125,305	143,613	142,945
Maryland	775	1,615	2,606	3,290
Ohio	34,000	55,351	46,770	46,205
Pennsylvania	65,500	80,491	84,137	83,225
Tennessee	5,000	8,237	8,206	10,320
Virginia	29,000	35,016	35,510	37,850
West Virginia	119,500	144,072	109,283	95,405
Illinois	46,000	65,119	59,537	53,880
Indiana	15,100	22,263	25,124	27,995
Total east	391,575	558,029	537,430	522,335
Arkansas	405	268	488	550
Colorado	3,630	6,025	8,219	1,190
Iowa	1,050	987	622	525
Kansas	800	1,627	479	630
Missouri	2,825	4,447	5,638	6,625
Montana	300	3,447	22,054	29,320
New Mexico	260	7,361	8,785	11,255
North Dakota	2,500	5,639	8,515	12,165
Oklahoma	1,300	2,427	2,872	5,345
Utah	4,925	4,733	6,961	9,240
Washington	225	37	3,743	5,055
Wyoming	1,950	7,222	23,804	44,500
Total west	20,170	44,220	92,180	137,130
Other[a]	755	681	22,041[a]	29,090[a]
Total U.S.	413,000	602,932	651,651	688,555

Source: National Coal Association, *Bituminous Coal Data,* various years.

[a] For 1975, Arizona 6,986, Alaska 766, Georgia 74, Texas 14,215. For 1977, 11,475, 665, 185, 16,765, respectively.

These trends are illustrated in table 1.3. The percentage of total production accounted for by underground mining declined throughout the east during the 1950s. In the 1960s this trend began to bottom out. In the 1970s the decline had stopped in several important eastern coal-producing states.

With these structural changes has come an enormous increase in regulation of the coal industry. Legislation has been passed controlling pollution both at the point of production and consumption. The health and safety of miners has become an issue receiving a great deal of public and legislative attention. The laws have been controversial and are continually being changed.

Pollution at the Mining Site
Strip mining is particularly disruptive to the environment. In hilly areas, such as Appalachia, great damage can be done to the topography. Land that is not reclaimed remains an eyesore. More important, damaged land is a source of contamination to rivers and groundwater. Rain reacts with sulfur in coal refuse piles and overburden spoils to produce sulfuric acid that finds its way into rivers and groundwater. In arid regions west of the Mississippi River, water contamination is not a problem. However, once the arid land is disturbed, it is not clear whether original vegetation can be restored. Furthermore in western areas coal seams are often aquifers, so removal of coal can jeopardize groundwater supplies.

Deep mining also has associated environmental costs. Surface subsidence can be a problem. The removal of coal seams removes support for overlying structures. The surface collapses if pillars of coal adequate to support the surface are not left in place. Acid runoff from abandoned mines pollutes rivers and streams. The disposal of coal refuse also creates environmental problems. While these problems are serious, they are on

Table 1.3 Percentage of output accounted for by deep mining for selected eastern states, 1950 to 1977

	1950	1960	1970	1975	1976	1977
Pennsylvania	75.0	67.4	68.8	53.0	51.0	46.0
Ohio	39.7	27.1	32.7	33.0	35.7	30.0
West Virginia	91.0	91.8	80.8	80.9	80.5	78.0
Kentucky	82.2	66.5	50.0	45.7	44.8	44.0
Illinois	68.7	50.7	49.3	53.5	53.2	55.0
Total U.S.	76.0	68.6	56.2	45.2	43.3	38.3

Source: U.S. Bureau of Mines, *Mineral Industry Surveys,* various years.

the whole less severe than environmental disruption caused by strip mining. However, deep mining is a dirty and relatively dangerous occupation. Questions of health and safety are thus of enormous importance.

Health and Safety
Underground coal mining is one of the most hazardous occupations in the United States.[4] Long-term disability is of particular concern. Pneumoconiosis, also known as black lung disease, leads to long-term disability and death. The Coal Mine Health and Safety Act was passed in 1969 to address these problems. The act sets standards for underground safety. Since passage of the act, fatal accidents have diminished. Nonfatal accidents declined through 1975 but then began to rise again. Productivity has also declined, and the act has been blamed for this decline. It is not yet clear how much of the decline is attributable to the legislation, although most would maintain there is at least some connection; see Baker et al. [2].[5]

Pollution at the Consumption Point
Coal also presents problems at the point of use. Coal burning results in emissions of particulate matter and sulfur dioxide. Both these forms of pollution are associated with respiratory ailments. Particulates can be controlled by electrostatic precipitators at some cost. The technology is well known and has been used for a long time. Sulfur presents more of a problem. There are basically two ways of reducing sulfur emissions: burning a coal lower in sulfur or using a stack gas scrubbing device. Coal contains sulfur organically linked to the coal, as well as sulfur present in pyritic particles not chemically bonded to the coal. The latter can be removed by mechanical washing devices prior to coal combustion. The former cannot be so removed. The percentage of sulfur that can be removed by washing varies from coal seam to seam. Since washing is not a complete solution to sulfur removal, there is great interest in processes that remove sulfur from the effluent gases during combustion.

The most important of the processes for removing sulfur from the effluent stream is stack gas scrubbing. There are several competing technologies. The most widely used technology uses limestone to scrub the effluent gases. Gases are passed through limestone which reacts with the sulfur in the gas to produce a calcium sulfate sludge. However, this sludge presents a disposal problem. Newer technologies are regenerative in that the limestone is separated out and used again, lessening the disposal problem.

There has been a great deal of controversy surrounding the use of scrubbers. Utilities have continued to maintain that they are costly and unreliable. The Environmental Protection Agency has taken a much more optimistic view and is mandating their use in new power plants.

The Interaction of Environmental Goals

The reduction of sulfur emissions conflicts with the desire to minimize damage from strip mining, for the following reason. The large incremental supply of low-sulfur coal is found in the western United States. However, that coal will be produced by strip mining. One goal trades off against the other. Sulfur pollution can be reduced at the expense of greater strip mining. Scrubbing devices, if they were low in cost, would make that trade-off easy. We could lower sulfur pollution and rely on eastern deep mining. Unfortunately scrubbing is costly.

The policies that the nation ultimately chooses will have to trade off among these various goals and the goal of substituting coal for imported fuel. The more coal used, the more the associated pollutants. The difficulty in choosing arises because we have a poor notion of the costs and the benefits. The costs of sulfur pollution in terms of disease, for example, are still not totally understood. How do we weigh the esthetic damage of strip mining? How much will national security benefit by substituting coal for oil in, say, industrial boilers? This study does not propose to answer these questions. Rather we take some goal as given and ask: What will it cost to reach that goal? We postulate, for example, a moratorium on strip-mining leases in the west and ask what that does to the cost of coal and of electricity. Then we do the same with sulfur pollution regulation, and so forth. In this way we can determine how costly it would be to achieve a goal. The reader is left to decide whether the goal is worth the cost.

1.2 The State of Regulation

In 1977 Congress passed and the President signed the Strip-Mine Reclamation Act. The bill calls for restoration that permits a use of the land at least as valuable as the use prior to mining. The original contour of the land must, where possible, be restored, and the mined land revegetated. The bill also specifies use of mining procedures that facilitate restoration and minimize damage to the land.[6]

Estimates of the cost of complying with these standards have ranged widely. Potential effects of this law are still unclear. In Appalachia, where mining takes place on hilly slopes, the incremental cost of restoring original contours will be substantial. In flat terrains with adequate rainfall the cost will be less. Without adequate rainfall revegetation is problematical. Since many states already have strip-mining laws as stringent as the new law (for example, Pennsylvania), some claim that the incremental effects will be small.[7]

Strip-mine reclamation requirements will have the least effect in the western part of the United States. There the coal seams are thick and tons of coal removed per acre are large. Therefore reclamation costs per ton of coal is apt to be small. However, because the western coal-producing regions are arid, there is serious doubt about the ultimate quality of revegetation.

The air pollution situation is complicated. In essence there are two types of air pollution standards. The first relates to ambient air quality. Each state has established standards in order to achieve the ambient air quality specified by federal regulation. These standards have resulted in implementation plans that vary from state to state. A typical regulation stipulates a limit on the sulfur content of a fuel or on the emissions of sulfur dioxide. In addition Congress established nondegradation standards for areas where the air is already cleaner than that required by the basic ambient air quality standards.

There are three different levels of nondegradation standards that specify allowable increases in pollution. The strictest standards apply to areas such as national parks. Other areas are initially classified in the less strict second or third categories but can be changed on petition of the governor of the state. In areas not in compliance with applicable ambient air standards, no new facilities are allowed unless plans for attaining compliance are presented or there is a compelling need for the new facility.

The second type of standard regulates emissions from pollution sources constructed or planned after 1971. It is called New Source Performance Standards (NSPS). This standard was set at 1.2 lb of sulfur dioxide per million Btu burned. Such a standard could be met by burning low-sulfur coal or by any combination of low-sulfur coal and mechanical or chemical cleaning devices.

The NSPS was modified by the 1977 amendments to the Clean Air Act. The new amendments called for the use of the Best Available Control Technology (BACT) for the removal of sulfur from coal. The amendments specifically stated that burning low-sulfur coal was not an acceptable sulfur reduction technique. In accordance with those amendments the administrator of the EPA on September 12, 1978, promulgated a new set of New Source Performance Standards. The preliminary standards required all major new coal-burning installations to reduce sulfur emissions by 85 percent from what would be emitted by the raw coal alone. Coal having an initial sulfur content of less than 0.2 lb SO_2 per million Btu was exempt. All new plants, in addition to achieving this reduction, were required to emit no more than 1.2 lb SO_2 per million Btu burned. If these standards had been adopted, it would have meant the scrubbing of all coal since 0.2 lb SO_2 per million Btu is so low a level of sulfur content that almost no coal in the United States will be below or at that floor.

Critics of the NSPS charged that these regulations would be costly. In high-sulfur coal-producing areas they might even increase total sulfur emitted, because scrubbing high-sulfur coal could emit more sulfur than using low-sulfur coal alone. The Department of Energy suggested a higher floor, so a larger amount of western coal could be burned without scrubbers. In addition a sliding scale for sulfur removal was proposed. This would require less percentage reduction as the raw sulfur content was reduced.

The final standards allow some partial scrubbing. All coal must be scrubbed. The minimum level of scrubbing is 70 percent removal. That means, regardless of sulfur content, at least 70 percent of SO_2 emissions must be removed by scrubbers. If emissions are greater than 0.6 lb SO_2 after scrubbing, the level of scrubbing efficiency must be 90 percent. If emissions are less than 0.6 lb SO_2 per million Btu after scrubbing, the utility may choose any level of scrubbing between 70 and 90 percent.

A large part of the impetus for modifying the NSPS was the desire to protect eastern coal markets. This desire is manifest at another level. Many states, particularly Ohio and Illinois, are seeing an increase in the use of low-sulfur coal imported from outside their states in order to satisfy State Implementation Plans (SIP). The plans require a reduction in emissions in existing as well as new sources. The increase in the import of out-of-state coal threatens local high-sulfur coal production. Ohio and Illinois are both seeking a modification of the standards that will permit them to use local coal. The EPA administrator can do this under the mandate of the 1977 amendments to the Clean Air Act. In sum the battle is being fought over environmental regulations, and the goals relate as much to where coal comes from as to how much sulfur will be emitted.

1.3 Nuclear Power and Coal

Coal and nuclear power are the main competitors for the new base-load capacity in the electric utility sector. Therefore developments in the nuclear area have an important impact on the coal industry. At the present time the nuclear power industry is in trouble in this country. Even before the events at the Three Mile Island station, California and Wisconsin had passed laws that limit new nuclear construction until the development of a national policy on radioactive waste disposal. The governor of New York is calling for similar legislation in his state. The Nuclear Regulatory Commission recently withdrew its approval of the summary of the major study purporting to show that nuclear power is a safe technology. Currently there is a moratorium on new construction permits.

Other regulatory uncertainties are affecting decisions to build new nuclear plants. To the extent that uncertainties raise the cost of nuclear plants, they affect the demand for coal. Thus the costs of any environmental policy with respect to coal are affected by the cost and availability of nuclear power.

1.4 The Scramble for Rents

The increase in demand for low-sulfur coal due to regulations coupled with the difficulties for nuclear power has led to an increase in the market power of those who can control western low-sulfur coal supply. The industry itself is competitive, so that potential power lies only with those who control key inputs and the state legislatures, which can levy taxes.

The western states are likely to take advantage of the situation by raising their severance taxes. Montana has already raised its tax to a level (30 percent) that is being challenged by utilities and coal producers as unconstitutiona.[8] Wyoming, with a 16.5 percent tax, is considering an additional 5 percent levy. The Internal Revenue Service has proposed a national coal tax. Other western states can be expected to follow in an attempt to capture the rents created by government regulation; see Alt and Zimmerman [1]. This is an issue both of efficiency and income distribution.

The railroads are also reacting to the increased demand for coal. In recent filings with the Interstate Commerce Commission, railroads have asked for rate increases for coal higher than those for general freight. The ICC has thus far refused. But for new rates, particularly new western shipments, the picture is changing. The western railroads claim that they need rates much higher than previous levels because of the new investment necessary for new shipments. The utilities claim that these higher rates are not justified by costs but represent an attempt to take advantage of the increased demand for coal. Several judicial cases have risen along these lines, the most publicized being the San Antonio case. The Municipal Electric Utility of San Antonio, on the basis of a planning rate quoted to them by the Burlington Northern Railway, contracted to purchase coal from Wyoming. The original contract was signed in 1973, and stipulated delivery beginning in 1974. When it came time to begin shipments, Burlington Northern quoted a rate 30 percent higher.[9] The utility claimed the railroad was taking advantage of the changed market circumstances, while the railroad claimed the rate was necessitated by its high costs. Other challenges to rates have been made by Southern Company, Virginia Electric Power Company. In response to these changes, the ICC has initiated proceedings to determine maximum reasonable rates for western coal. Furthermore deregulation of the railroads will increase their ability to capture rents.

The coal industry is being called upon to play a larger role in supplying U.S. energy needs at a time of increasing environmental concern and regulation. It is clear that trade-offs will have to be made. These may be made explicitly, or more likely fragmented decision making will lead us into policies that never explicitly consider the trade-offs and interactions.

1.5 The Method of Analysis

The object of the following chapters is analysis, not prediction. The energy picture is changing rapidly. Predicting how the industry will evolve to the year 2000 is a task not likely to be fruitful. Rather we use the models described in chapter 2 to test sensitivity to various changes affecting the coal industry. We examine relative changes to key variables. In many cases we purposely choose optimistic or pessimistic assumptions in order to bound our analysis. If under optimistic assumptions costs associated with changes in a particular variable are high, then we can be confident that we have isolated an important variable. The goal is understanding what drives the industry and what possible changes are likely to be important.[10] In the last chapter we test the sensitivity of our conclusions to recent developments in energy markets.

We also are not concerned with specific policies. We focus rather on the broad trade-offs the United States must make. The themes of this study are environmental regulation versus oil imports, sulfur pollution versus strip mining, and so on. We do not examine, except in passing, detailed implementation of policies. In this way it is hoped the relevance of the study will outlast the rapidly changing policy actions of the government.

References

1. Alt, C., and M. B. Zimmerman, "The Western Coal Tax Cartel." MIT Energy Laboratory working paper (forthcoming).

2. Baker, J. G., et al. *Determinants of Coal Mine Labor Productivity Change.* Final report to the U.S. Department of Energy. Contract no. DOE-AC05-760R00033.

3. Data Resources, Inc., "The DRI/Zimmerman Coal Model Documentation," June 1978.

4. Gordon, R. L. "The Hobbling of Coal: Policy and Regulatory." *Science,* 200 (April 14, 1978): 153–158.

5. ICF, Inc. *Energy and Economic Impacts of H.R. 13950.* Report to the Council on Environmental Quality and the Environmental Protection Agency. Contract no. EQ6AC016, February 1, 1977.

6. U.S., Department of Energy, Energy Information Administration. *Analysis of Proposed U.S. Department of Energy Regulations Implementing the Power Plant and Industrial Fuel Use Act.* DOE/EIA-0212/21. Washington, D.C.: Government Printing Office, November 1978.

7. Zimmerman, M. B., and R. P. Ellis. "What Happened to Nuclear Power?" MIT Energy Laboratory working paper no. MIT-EL 80-002WP, January 1980.

2   The Models

A model approximates a complex system. The goal here is to develop a model for policy analysis. In building such a model, choices must be made between increasing complexity that more closely parallel actual rules and regulations and the intractability and computational cost that such complexity entails.

The coal model attempts to capture a highly complex industry and regulatory regime. The model structure attempts to capture the heart of environmental regulation, while remaining simple enough to reveal the basic working of the industry. The four model components are the coal supply model, the transport cost model, the demand model, and the linear programming link between the models. Parts of the supply and transport models, as well as the demand model, have been documented; see Zimmerman [24], [25], [26] and Baughman et al. [2].[1] The goal here is not to repeat that documentation but to present an overview of how the models are estimated, constructed and linked together.

2.1 Model Overview

Figure 2.1 provides a schematic overview of the coal policy simulation model. Given an initial set of prices, the initial demand for coal is determined by demand region. These demands are passed to the linear program as constraints. The supply model establishes the cost of producing coal as a function of output by sulfur content and region. The transport model determines the cost of transporting coal from mine mouth to demand center. The linear program then minimizes the cost of mining and transporting the coal necessary to meet the regional demands, subject to limitations on sulfur emissions in each region.

The solution of the linear program yields output by region and sulfur content, flows of coal from supply region to demand region, and the delivered price of coal to each region. The delivered prices are passed back to the demand model to estimate a new set of regional demands. The coal

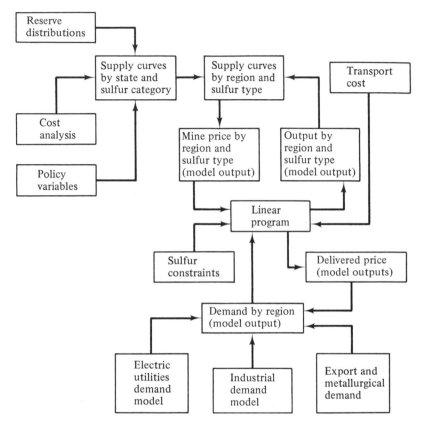

Figure 2.1 Structure of the coal policy model

outputs are fed back to the supply model to establish a new set of supply prices, and the process is repeated. In theory one could iterate until the solution converges to a set of prices consistent with instantaneous equilibrium in supply and demand. In practice we run the models recursively. The demand in year t is a function of price in year $t - 1$. This considerably reduces the computational complexity of the model without introducing any serious distortion. In fact the demand changes in response to price changes take many years to make a complete adjustment. In any given year the demand is quite inelastic with respect to prices.

The linear program is solved each year. In any given year, however, the output affects coal prices throughout the entire simulation period. The essence of the supply model is the cumulative cost function. This function estimates how costs increase as output cumulates over time. In each year the cost of producing any given amount of coal is updated to reflect cumulative output through that year.

Levels of Aggregation and Model Integration

The integration of models provides an interesting set of challenges. A model built to meet specific purposes often does not lend itself easily to being combined with other models built to analyze different questions. This was the case with the models used here. The supply model emphasizes aspects of supply and provides rich detail with respect to those aspects, whereas the demand model was built to analyze other issues and was not structured to deal with the richness of the supply side. Combining models can lead to computational complexity and expense that makes flexible policy analyses on a finite budget impossible.[2] The solution to this problem is to find the appropriate levels of aggregation so model interaction is feasible, while maintaining a level of detail consistent with the aggregate solution.

Regional Aggregation of Supply

The locational aspects of the coal industry are important. The regional distribution of income lies behind many of the significant policy issues. Attitudes toward coal development in individual states will affect the industry at a national level. There is a need therefore for regional and even state detail in the model. Regulations impact differently on strip mining and on deep mining, and therefore disaggregation of supply between mining types is also necessary. As sulfur pollution regulations change, so will the demand pattern. Regions producing low-sulfur coal will benefit as regulations on allowable emissions become tighter. Thus we also need disaggregation by sulfur content.

Allowing for this disaggregation yields 2 mining types, 18 states, and 10 levels of sulfur content, for a total of 360 supply curves in any year. This is computationally an unmanageable number to integrate with the demand analyses. We therefore sum the strip and deep cumulative cost curves and then the state supply functions to create six supply regions.[3] Figure 2.2 is a map of the regions (the supply regions are given in Roman numerals), and table 2.1 defines the regions according to states and the coal districts established by the Bituminous Coal Act of 1937. To reduce computational difficulties further, we aggregate sulfur categories to eight intervals, given in table 2.2.

This aggregation results in six regional supply functions for eight sulfur categories, for a total of forty-eight supply curves. The linear program

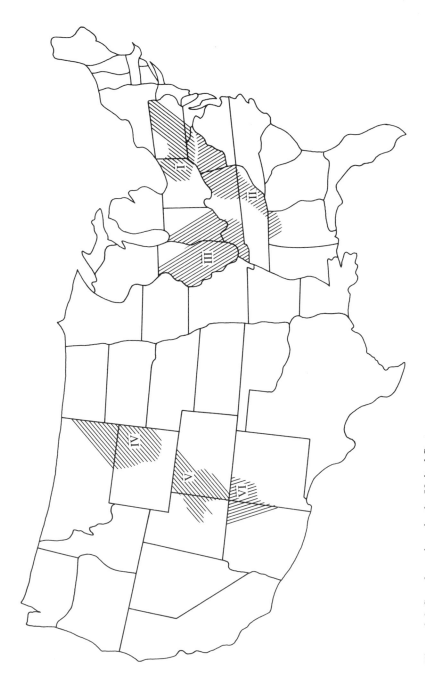

Figure 2.2 Supply regions in the United States

Table 2.1 Definition of coal regions

Coal model	States	Districts
1 Northern Appalachia	Pennsylvania, northern West Virginia, Ohio, Maryland	1, 2, 3, 4, 6
2 Southern Appalachia	Southern West Virginia Virginia, East Kentucky, Tennessee, Alabama	7, 8, 13
3 Midwest	Illinois, Indiana, West Kentucky	9, 10, 11
4 Powder River Basin	Montana, Wyoming	19, 22
5 Utah-Colorado	Utah, Colorado	16, 17, (excluding New Mexico, 20)
6 Arizona-New Mexico	Arizona, New Mexico	18, part of 17

Table 2.2 Sulfur categories used in model

Category	Sulfur limits
1	0.00–0.64
2	0.65–1.04
3	1.05–1.44
4	1.45–1.84
5	1.85–2.24
6	2.25–2.64
7	2.65–3.04
8	3.05 +

link deals only with these forty-eight curves. After the program solves for production by region and sulfur category, we can determine the split among states and mining type by using the aggregation process in reverse. All that is needed is the price of coal of a particular type in a particular region. The coal supply model aggregates all coal of a given sulfur content in the states within a region at a given price. In reverse, it is simply necessary to quote the price for that coal in the region and ask how much is available in each state. The state detail is not used in the regional model yet is available, if needed, for policy analysis.[4]

Regional Aggregation of Demand

Demand is calculated on a regional basis. In the demand model we have adapted, there are nine demand regions, based on the U.S. census regions.

But the number is too small from the standpoint of coal policy. The western regions are extremely large and not likely to behave as single units with respect to coal choice. We have therefore disaggregated the nine regions in the demand model into twelve regions for interface with the supply model. The twelve regions are shown in the map of figure 2.3. Census region 4 was divided into demand regions 4 and 10. Census region 8 becomes demand regions 8, 11, and 12.

The disaggregation is performed according to the percentage historical composition of each region. Once coal allocations and delivered prices are determined for the twelve regions, they are reaggregated with the same weights, and nine regional coal prices are fed back to the demand model.[5]

Sulfur Aggregation

In the demand model electric utilities choose among plant types, coal, nuclear, oil, gas, among others, on the basis of cost. The present value of fuel expenditures is added to capital expenditures to determine the least-cost capacity addition. We have introduced into the model the choice between high-sulfur coal plants that use scrubbers to meet pollution regulations and plants without scrubbers that use low-sulfur coal. There are, however, eight different sulfur levels, and it would be prohibitively complicated to force the demand model to consider all possible combinations. The solution adopted is to aggregate coal demand in each region into two sulfur categories: a high-sulfur and a low-sulfur category. The composition of these two categories is determined endogenously in the model.

There is a demand for coal that meets sulfur pollution requirements without stack gas scrubbing devices and demand for coal that meets standards only with stack gas scrubbing. Plants without scrubbers must meet applicable emission limits by using low-sulfur coals. The second category, coal used in plants with scrubbers, also must meet emission standards. Since scrubbers are assumed to be at least 85 percent effective in reducing emissions, the second category can in effect meet standards burning any coal. Thus its sulfur constraint is never binding. The linear program minimizes the cost of satisfying these two separate categories of demand by taking into account the appropriate sulfur constraint. The program chooses mixtures of coal that satisfy the sulfur constraints. For plants without scrubbers this is a mixture of low-sulfur coals. For plants with scrubbers it is a mixture of high-sulfur coals. The average price of the mixture in each category is the price of the combination of coals that satisfies the applicable sulfur constraint at least cost. In this way we allow mixing of coals, yet need pass only two prices to the demand model. The demand

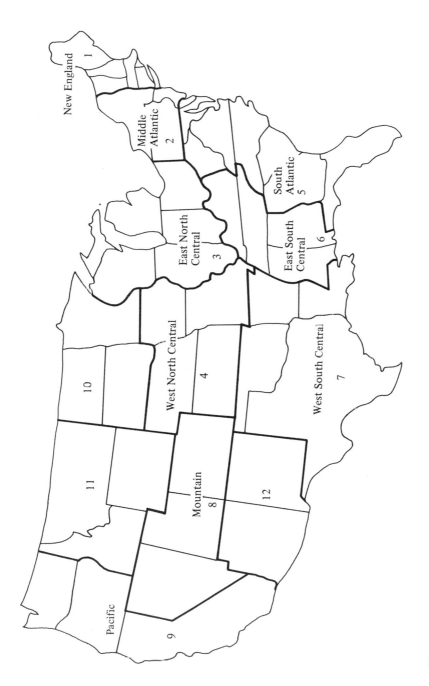

Figure 2.3 Demand regions in the United States

model uses these prices of high- and low- sulfur coal to determine whether to build plants with or without scrubbers. The distinction between low- and high-sulfur coal is created by the sulfur constraint. This constraint in turn reflects the legislation. Without a sulfur constraint there would be only a single price of coal, and the demand model would never choose scrubbers.

2.2 The Coal Supply Model

The coal supply model is a disaggregated model estimated specifically to perform policy analysis. Mineral industries present a unique problem to economists attempting to model supply response. Economic theory predicts that lowest-cost deposits are exploited first. As these deposits are exhausted, the more difficult to mine, and therefore more costly, deposits are exploited. The structure of the industry is continually changing due to this depletion phenomenon. Parameter estimates that capture past experience are not a guide to future behavior, because mining conditions change over time.

Depletion is only part of the problem of structural change in the coal industry. Regulatory policies are continually changing the structure of the industry. Policies toward sulfur pollution affect the kind of coal that can be burned and the premia paid to higher-quality coals. On the supply side regulatory policies have profound impacts on the cost and availability of coal. The Health and Safety Act has had a large impact on the costs of underground mining; the new strip-mining law has substantially raised the costs of mining on steep slopes; and leasing policy has reduced the availability of coal in the west. The rules change with each session of Congress. Since supply response estimated under previous rules is no longer a good prediction about the future, the problem is how to predict an industry structure that is experiencing constant change.

The coal supply model attempts to deal with this problem. The basis of the supply model is a cumulative cost function; see Hotelling [7].[6] This function yields long-run marginal cost as a function of cumulative output. It captures the effect of depletion. Structural change occurring through policy action is handled by the various levels of aggregation. Sulfur regulations, for example, change the demand for coals having various sulfur levels. Since the supply functions are estimated separately by sulfur content, we can examine the effect of these regulations on price and output in each region. We can examine the effect of tax policies because we have estimated supply functions for each state.

There are three main steps to the estimation process. In the first step a cost function is estimated that explicitly accounts for the impact of geology on the cost of production. In the second, the distribution of coal in the ground according to the geological characteristics of the deposit is established. In the final step, the cost function is combined with the reserve distribution to yield the cumulative cost function.

Coal Mining Technology

The cost function estimated is based on an engineering model of coal production. There are two primary techniques for producing coal. Underground methods are used for coal seams lying deep underground (usually more than 150 feet). Openings to the seam are constructed, and mining takes place beneath the surface. When the coal seams lie close to the surface, the earth and rock (overburden) above the seam are removed, and the coal is excavated at the surface.

Deep-Mining Technology

The dominant technique for mining coal underground in the United States, and the one upon which the following estimation is based, is continuous mining. In continuous mining a large machine rips the coal from the seam and loads it onto shuttle cars in one continuous operation. The shuttle cars transfer the coal to a central transport network for removal to the surface. A mining machine, two shuttle cars, and a complement of miners comprise a mining section. A mine consists of a number of sections, each working independently but sharing a common haulage system, a ventilation system, and a set of openings to the seam that provide a means for removing coal and access for miners and supplies. Noncontinuous techniques are also used, but at current factor prices the large new mine (the mine that determines the cost of coal at the margin) will use the continuous mining method. Within a mining section, the ratio of capital to labor is relatively fixed.[7]

Surface-Mining Technology

In surface mining the overburden is removed with large earth-moving equipment. After the seam is exposed, coal is removed, using smaller shovels, and loaded onto trucks. After coal removal the overburden is returned to the pit and the ground is recontoured and reclaimed. The equipment used in the United States for this type of mining varies from region to region. In the west and midwest, where relatively large parcels of reserves lie underneath flat or gently rolling terrain, large draglines and

shovels with capacities of up to 300 cu yd are used, whereas in the hills of
Appalachia smaller equipment is the norm, because hilly terrain forces
operations to take place on narrow hillsides which, with frequent moving
of machines, can be accomplished only with smaller, more mobile
machinery.

Surface mining is more capital intensive than deep mining, in part
because the size of draglines and shovels has increased dramatically over
time. One operator with one machine can move a large amount of
material in a machine cycle. In the last five years large draglines have
come to dominate the large-scale surface-mining operation, because of
their great flexibility.[8] The estimation below is based on dragline operation, although this is not an important factor; consideration of shovels
would lead to only minor changes.

The Estimation Procedure
When reading a coal-mining manual, one is impressed by the multiplicity
of geological factors that affect costs:

Natural conditions involve roof, floor, grades, water, methane and the
height of the seam . . . . In addition to these normal conditions, there
are, in some mines, rolls in the roof or floor, and clay veins of generally
short horizontal distance that intersect the coal seam. All these must be
taken into account.

It is possible for an experienced engineer to examine previous conditions
of the sections and the immediate area of the section and assess proper
penalties. As an example, if the roof is poor, production is reduced by as
much as 15 percent of the available face working time. If the floor is soft,
fine clay and water is present, the production handicap could be as much
as 15 percent. If a great deal of methane is being liberated, so that it is
necessary to stop the equipment until the gas has been bled off, this delay
could run as high as 10 percent. Fortunately, only a few mines in the
United States have such severe conditions. The same remarks apply to all
the other natural conditions.[9]

The following is a partial list of factors taken from an authoritative mining manual that must be considered when planning a surface mine:[10]

1. total coal reserves and general arrangements of deposit,
2. annual capacity,
3. nature of terrain,
4. maximum depth,
5. nature of overburden,
6. type of overburden.

Unfortunately data are not generally available on all these cost-
determining variables.

Step 1: The Cost Function   Ideally the following cost function would be estimated:

$$c = (G_1, G_2, G_3, \ldots, G_n, Q),$$

where G_i = the n different appropriate geological characteristics and Q = output of the mine. For deep mining the only data available are on seam thicknesses. Assuming seam thickness is independent of other factors, the following relationship is estimated:

$$c = (Th, Q) + \epsilon,$$

where Th = seam thickness and ϵ = disturbance term, reflecting other natural mining conditions. In a statistical sense ϵ represents the variability in cost that cannot be explained by seam thickness. It represents collectively all other natural conditions in the mine. It is in effect an index of the favorability of unobserved mining conditions. The more important the influence and the more variable the incidence of these factors, the greater will be the variability of ϵ.

Similarly, assuming overburden ratios are independent of other mining conditions, a strip-mining cost function is estimated as

$$c = (R, Q) + \eta,$$

where R is the ratio of feet of overburden to feet of coal seam and η is a disturbance term, reflecting the unobservable mining conditions.

Since no data are available on cost by mine, cost functions cannot be estimated directly. Instead an indirect procedure is used: the production process within the mine is modeled as the sum of the operations of individual production units. In deep mining these are the mining sections. In strip mining the relevant unit is the size of the dragline used to remove the overburden. We start by estimating the relationship between the productivity of the relevant producing units that comprise a mine and the output and geological factors. Formally

$$\frac{Q}{u} = f(Q, G_i) + \epsilon, \tag{2.1}$$

where

Q = output of the mine,
u = the number of producing units within the mine,
G_i = observable geological characteristics,
ϵ = unobservable geological characteristics.

We then assume that capital expenditures and expenditures for labor and supplies are functions of the number of producing units:

$$E_j = g(u), \quad j = 1 \ldots N, \tag{2.2}$$

where E_j refers to the class of expenditures such as labor, capital, material, and supplies. We obtain a cost function by combining equations (2.1) and (2.2). Substituting (2.1) into (2.2) yields E_j as a function of Q, G_i, and ϵ. The cost E_j is a function of output and observable and unobservable geological characteristics.

Deep-Mining Cost Functions
The productivity of an underground mining section is related to the thickness of the seam, Th, the number of sections in the mine, S, and the number of openings (drift, shaft, and slope) to the seam, Op. Since many unobservable geological conditions also affect mining productivity, a statistical relationship is estimated that explicitly measures the importance of these unobserved conditions on output per section. The following equation is estimated:

$$\frac{Q}{S} = A Th^\gamma S^\beta Op^\alpha \epsilon, \tag{2.3a}$$

where

A, γ, β, α = constants,
$\quad \epsilon$ = disturbance reflecting the impact of unobserved mining conditions.

The multiplicative form of the equation expresses the interaction between natural mining conditions and seam thickness. By including the number of sections in the productivity equation, we capture the scale effect on productivity per unit. Larger mines will have greater logistics problems. The larger the mine, the greater will be the travel time to the working sections. Production per section should therefore decline as congestion takes its toll. The problems faced by larger mines can be mitigated by capital expenditures. As congestion effects increase costs, new openings to the seam can be constructed, lessening the difficulties. In other words, the observed scale effects reach a limit as new shafts are opened to the seam. To capture this effect, the number of openings to the seam is included as an explanatory variable in the productivity equation.

Equation (2.3a) is estimated using data on a sample of all deep mines producing more than 100,000 tons per year in 1975; see Zimmerman [24].[11] Economic theory signals a problem for this estimation. We know that only mines that were expected to be profitable were opened. There are potential mines that have not been opened. This means that observed mines with thin seams must be compensated by a more favorable set of other mining conditions. In other words, ϵ in the sample is correlated with the observed seam thickness. This is another manifestation of depletion and another difficulty in attempting to use past data to predict future behavior in a mineral industry.

The problem here is that the sample is biased. Economics has dictated that we will only observe combinations of seam thickness and ϵ that result in a productivity level great enough to make a profit. We therefore estimate equation (2.3a) by a maximum likelihood technique that takes explicit account of this truncation problem; see Hausman and Wise [6].[12]

There is one remaining econometric problem. The observed data on productivity per section are simply total production divided by numbers of production units. In large mines the shutdown of a single mining unit will not greatly affect either total production or measured productivity per section. In a small mine, however, a shutdown of one unit might represent a loss of half the total output. Observed productivity per section will have changed dramatically. In other words, the variance in observed productivity per section should be inversely related to the number of mining units. This is a problem of heteroskedasticity, and we account for this by weighting all variables by the square root of the number of sections.

The resulting maximum likelihood estimates are

$$\sqrt{S} \log \frac{Q}{S} = 0.7568\sqrt{S} + 1.1071 \ (\log Th)\sqrt{S} - 0.2185 \ (\log S)\sqrt{S}$$

(s.e.)   (0.4842)   (0.1205)   0.0594

(t-statistic)   (1.5630)   (9.1906)   (-3.6762)

(2.3b)

$$+ 0.0283 \ (\log Op)\sqrt{S},$$

(0.0655)

(0.4314)

S.E.R. $= 0.9799$,

$x^2 = 232.4285$,

$N = 244$.

The results indicate that seam thickness is important. A 1 percent decrease in seam thickness results in over a 1 percent decrease in production. However, the standard error of the regression (the estimate of the standard deviation of ϵ) indicates that seam thickness is not the only important geological variable. The unobserved mining characteristics are also important. Finally, the estimate of scale effects indicates that there are decreasing returns as the number of units increases.

Deep-Mining Expenditures

The estimation of the expenditure equation for deep mining is accomplished using engineering estimates prepared by the U.S. Bureau of Mines [19]. These estimates consider the equipment, labor, and supplies necessary for hypothetical mines. These estimates are for mines assumed to produce a given level of output in a seam either 48 or 72 inches thick under some unspecified set of mining conditions.

Using these estimates, the following equation was estimated for each class of expenditures:

$$E = \alpha + \beta S + \xi, \tag{2.4}$$

where

E = expenditure,
S = number of sections,
ξ = disturbance term.

Capital expenditure equations are estimated for three classes of capital equipment, subdivided by operating lifetime. These are shown in table 2.3. This breakdown is necessary because the correct treatment of depreciation requires adjustment for the life of the equipment. Capital expenditures are amortized over the life of the equipment and expressed as an annual capital cost per ton. The cost of capital used in amortizing the expenditures is based on a 10 percent real rate of return after tax, the sum-of-years form of depreciation, and allows for a 50 percent of gross profits depletion allowance, the 10 percent investment tax credit, and the corporate income tax.[13]

Associated with the direct expenditures are outlays for engineering, overhead, and various small construction tasks. The Bureau of Mines assumes these to be 28 percent of initial direct capital expenditure. Equations are also estimated for labor costs and for materials and supplies (see table 2.3). Labor costs reflect union scales as of 1975; they were adjusted to 1977 levels by multiplying by an index based on the recent United Mine Workers of America (UMW) contract. Capital costs were escalated by BLS indexes for mining equipment. To these costs were added an additional 15 percent of direct operating costs (labor plus materials and supplies). These fixed percentages also come from Bureau of Mines estimates. The relevant UMW per ton charges are also added.[14] The number of openings to the seam is taken as two, and equation (2.3b) is solved for S as a function of output and geological condition.[15] The resulting expression for S is substituted into equation (2.4), yielding a total cost equation. Total annualized cost for a mine is a function of output and geological conditions:

$$TC = a_0 + a_1 \frac{(Q)^{a_3}}{a_2 (Th)^{\gamma_\epsilon}},$$ (2.5)

where

$a_0 = \$3,914,764,$
$a_1 = \$2,122,480,$
$a_2 = 1597.6,$
$a_3 = 1.2796,$
$\gamma = 1.1071.$

Strip-Mining Cost Functions

The procedure followed when estimating the strip-mining cost functions is analogous to that used for estimating deep-mining cost functions. The analysis begins with the productivity of a mining unit, except that a unit is defined in a rather different way.

The dragline is the most commonly used equipment for overburden removal in large-scale strip mining. Over the past five years it has dominated new equipment purchases. Therefore the cost functions developed here are based on that technology. The size of draglines used in industry has increased steadily over time. Since draglines vary in size, the unit of analysis is not the dragline itself but its size. Productivity is defined as the output, or overburden removed, per unit of size. We take as the measure of the size the maximum usefulness factor, or *MUF*. This concept arises from engineering analysis; it was introduced by Rumfelt [13]. The MUF is the product of the bucket capacity in cubic yards and the radius of the machine, as shown in figure 2.4. In essence it is a measure of the work of which the dragline is capable.

The size of the dragline chosen depends upon the amount of overburden to be removed. That in turn depends upon the ratio of feet of overburden to feet of coal seam and the output of coal. If R represents the ratio of feet of overburden removed to feet of coal seam exposed, the total cubic yards of material to be removed is[16]

$0.89(RQ).$

Table 2.3 Underground-mining expenditures as a function of sections

Equation	Item	Constant
(11)	I_5 5-year capital	88380.5 (39528.5) (2.23587)
(12)	I_{10} 10-year capital	2193350.0 (254037.0) (8.6340)
(13)	I_{20} 20-year capital	597135.0 (36493.1) (16.3629)
(14)	I_T Total capital	2878870.0 (36493.1) (13.6973)
(15)	LAB Labor	407608.0 (61671.2) (6.59974)
(16)	SUP Supplies	-100713.0 (94557.6) (-2.12265)

Note: Numbers in parentheses are the standard error and t-statistic, respectively.

We then relate the size of dragline to the amount of overburden to be removed:

$$\sum MUF = A(0.98)^\alpha (RQ)^\alpha N^\beta \eta,$$

where

$\sum MUF$ = the total capacity of all machines,

$A, Z, \alpha, \beta,$ = constants,

R = overburden ratio = feet of overburden per foot of coal,

Q = annual output rate,

η = stochastic term reflecting unobserved geological characteristics, assumed lognormally distributed,

N = the number of machines in use.

The number of draglines in use, N, is added to capture the effect on productivity of using more than one machine. The equation that results from the estimates is

$$\log MUF = -0.446684 + 0.612306 (\log RQ)$$
$$\text{(s.e.)} \quad (2.31895) \quad (0.145663)$$
$$\text{(t-statistic)} \quad (-0.192623) \quad (4.20357) \tag{2.6}$$
$$+ 0.506967 (\log N)$$
$$(0.779134)$$
$$(2.21254)$$

Slope	R^2	$F(1/5)$	S.E.R.
25528.0 (3102.92) (8.22708)	0.9312	67.6843	46202.7
546842.0 (19941.0) (27.4224)	0.9934	751.994	296930.0
53870.7 (2864.65) (18.8053)	0.9861	353.640	42654.8
626241.0 (16498.6) (37.9571)	0.9965	1440.75	245666.0
509088.0 (4848.14) (105.007)	0.9995	11026.5	72189.2
352102.0 (7422.6) (47.4365)	0.9978	2250.23	110523.0

$R' = 0.8595,$
$F(\frac{2}{7}) = 21.4089,$
S.E.R. $= 0.287762.$

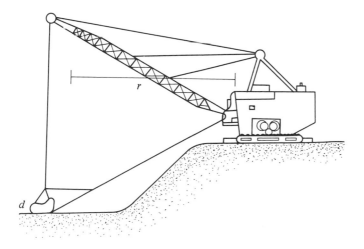

Figure 2.4 Dragline used for overburden removal in large-scale strip mining

The statistical results indicate that one large dragline is more productive than two smaller draglines with an equal capacity. This is consistent with the recent movement to larger machines. In the cost curve construction we assume $N = 1$.

Strip-Mining Expenditures
Following the procedure used for deep mining, expenditures in strip mining are related to the number of producing units, in this case the size of the dragline. The size of dragline is a function of the amount of overburden to be removed. However, there will also be expenditures solely dependent on the amount of coal produced. Therefore each class of expenditure is related to both the size of dragline and the annual production of coal. An equation of the following form is estimated:

$$E = \alpha + \beta(MUF) + \gamma Q, \tag{2.7}$$

where Q is annual output in tons. The results are shown in table 2.4.

The data are again taken from engineering estimates of the Bureau of Mines [17].[17] All indirect expenditures are treated in the same manner as those for deep mines. The cost of the dragline is estimated as a function only of its size. The data used for the dragline cost equation come from dragline manufacturers' list prices and dragline specifications.

The capital expenditure equations are converted to an annual required revenue stream, as was done for deep mining. To these capital costs are added the materials and labor costs. Again, allowances are made for indirect costs, fringe benefits, and so on. The resulting total cost equation is

$$TC = 3,170,223 + 467,262MUF + 0.96Q. \tag{2.8}$$

Substituting (2.6) into (2.8), and setting $N = 1$, yields

$$TC = 3,170,233 + 299(QR)^{0.6123}\eta + 0.96Q. \tag{2.9}$$

Long-Run Marginal Cost
In the long run the marginal cost of producing coal is the minimum average cost of a new mine. The task is thus to establish the minimum efficient scale of a mine and then evaluate the average cost at that point. We obtain the equation for minimum average cost for deep mining by dividing equation (2.5) by Q, differentiating the resulting average cost equation with respect to output, setting the result equal to zero, and solv-

ing for the output rate that minimizes average cost. The minimum average cost is obtained by substituting the minimum average cost output into the average cost equation. This procedure yields the following equation for the minimum average cost of deep mining:

$$AC^* = \frac{2567}{Th^{1.1071}\epsilon}. \tag{2.10a}$$

Equation (2.9) demonstrates increasing returns to scale for strip mining. Average costs are declining over the range in the sample. There are, however, constraints operating in the short run. Transportation of large components presents problems. Welding and other technologies need to be developed; thus at any point in time there are limits to the economies of scale. There are also nontechnological limits to scale economies. In hilly terrain, for example, small mobile draglines are necessary. We estimate a constraint on size in the following way. Equation (2.6) gives a relationship between R, Q, η, and machine size. New mines opening in an area will push to the operative constraint in terms of machine size. If we could observe R, Q, and η, we would know the constraint. Unfortunately η cannot be observed. Since η is assumed lognormal, we estimate the constraint in each region by substituting the geometric mean of R, Q for new mines into (2.6) with a value of 1 for η; see the *Keystone Coal Manual* [9] and Nielsen [10]. Substituting the constraint into (2.6) and dividing by Q, yields (2.10b).

West (Power River Basin)
$$AC^* = 0.52\,R\eta^{1.63317} + 0.96,$$

Midwest
$$AC^* = 0.67\,R\eta^{1.63317} + 0.96,$$

East
$$AC^* = 0.82\,R\eta^{1.63317} + 0.96.$$

Step 2: Reserves and cumulative costs. The long-run marginal cost curves are a function of Th, R, η, ϵ, and the estimated parameters. We need to know how deep coal is distributed among seams of various thicknesses and ϵ and also how strippable coal is distributed by overburden ratio and η.

The distributions of ϵ and η are assumed to be lognormal, and the parameters of those distributions are estimated in the process of estimating equations (2.3) and (2.6). The variance estimates are squares of the

Table 2.4 Strip-mine expenditures

Item	Constant	Coefficient of *MUF*
Dragline	158650.0	593.48
Cost	(702532.0)	(44.4945)
(*t*-statistic)	(2.25826)	(13.3383)
I_T	3107220.0	17.2718
(all capital)	142884.0	(5.72053)
other than	(21.7465)	(3.01926)
dragline)		
I_5	268374.0	− 12.0284
	110344.0	(4.41771)
	(2.43215)	(− 2.72274)
I_{10}	596687.0	1.72287
	(110344.0)	(1.69333)
	(14.1078)	(1.01745)
I_{20}	1762480.0	22.02644
	(107486.0)	(4.30334)
	(16.3973)	(5.11844)
I_{201}[a]	479674.0	5.55102
	(93412.2)	(3.73987)
	(5.13502)	(1.48428)
LAB	448156.0	− 11.7455
	(68894.9)	(4.40273)
	(6.50492)	(2.66778)
SUP	− 61161.9	62.0065
	(152040.0)	(9.71610)
	(− 0.402276)	(6.38183)

Note: Numbers in parentheses are the standard error and *t*-statistic, respectively.
[a] I_{201} represents capital ineligible for the investment tax credit.

standard errors of the respective productivity equations. The means of log ϵ and log η are assumed to be zero. All that remains to be estimated are the distributions of *Th* and *R*: the distribution of coal in the ground, according to seam thickness for deep mining and overburden ratio for strip mining.

The Distribution of *Th*

The U.S. Geological Survey provides coal reserve data in broad seam thickness intervals. They record coal lying in seams between 28 and 42 inches thick and seams greater than 42 inches thick. Considering the impact of seam thickness on cost, this is inadequate for a complete description of the cumulative cost function. The complete data are not available.

Coefficient of Q	R^2	F	S.E.R.
	0.9175	177.910	1885590.0
0.213019 (0.0577415) (3.68919)	0.9335	35.0953	222357.0
0.247652 (0.0445918) (5.55374)	0.8680	16.4385	171719.0
0.0112230 (0.0170920) (0.656623)	0.4681	2.20031	65819.6
−0.114750 (0.0434368) (−2.64176)	0.8461	13.7415	167271.0
0.0688944 (0.0377493) (1.802505)	0.7736	8.54032	145369.0
0.0427509 (0.0315682) (1.35424)	0.8727	17.1455	107454.0
0.228441 (0.0696658) (3.27910)	0.9753	98.8775	237134.0

Instead we use here an approximation that establishes a method for more accurate estimation as more data become available.

Data are available on the distribution of tons of coal in the ground, according to the thickness of the seam in which it lies, for Pike Country, Kentucky, a large and important coal-producing county in East Kentucky; see [12]. The object is to approximate this actual distribution with a well-known statistical distribution. The lognormal distribution has been used for similar purposes in studies of other minerals and proves useful here.

If lognormality is a good approximation, the points when plotted on log-probability paper will lie approximately along a straight line (see figure 2.5). Thus, for example, 50 percent of coal reserves lies in seams thicker than 42 inches. The fit is poorest in the tails, a common problem in work using the lognormal. However, the approximation appears adequate.

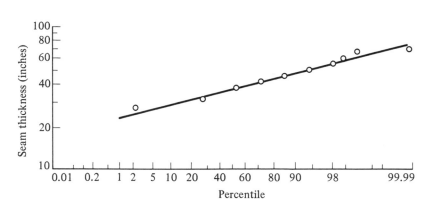

Figure 2.5 The distribution of coal reserves by seam thickness in Pike County, Kentucky (Source: Paul Weir Company, *Economic Study of Coal Reserves in Pike County,* January 1972)

The variance is measured from the data shown in figure 2.5. It is assumed that the lognormal distribution is applicable state by state and that the variance remains constant from state to state. The mean in each state is estimated separately, using Bureau of Mines data for coal lying in seams thicker and thinner than 42 inches. Those data yield a single point on the distribution function for each state and sulfur category. Recall that the distributions are assumed to be lognormal and that the variance is held constant at the value estimated for Pike County. We know what percent of total coal lies in seams greater than 42 inches. This cumulative percent corresponds to a point on the standard normal distribution, u_{42}:

$$u_{42} = \frac{log\ Th_{42} - \overline{log\ Th}}{\sigma_{(log\ Th)}}$$

where

$log\ Th_{42} =$ logs of seam thickness of 42 inches,

$\overline{log\ Th} =$ mean of the logs of seam thickness,

$\sigma_{(log\ Th)} =$ standard deviation of the log of seam thickness,

$u_{42} =$ the point of the standard normal distribution corresponding to log 42 inches;

or

$$\overline{log\ Th} = log\ Th_{42} - \sigma_{log\ Th}\,(u_{42}).$$

This expression can be solved for the mean of the log of seam thickness in each state. Thus, if 50 percent of the coal lies in seams greater than 42 inches, u_{42} is the 50 percent point on the standard normal, or zero.

Substituting the value of u_{42} allows us to solve for $\overline{\log Th}$. This procedure is repeated for each sulfur category, ranging from less than 0.4 percent sulfur to over 4.0 percent by weight. It is assumed that sulfur content is distributed independent of seam thickness.

Since tons of coal by the log of seam thickness are distributed lognormally, the distribution of tons of coal in the ground according to the log of the cost of mining is the sum of two normal distributions and is therefore itself lognormal.

The Distribution of R

When estimating the distribution of R, there is again a paucity of data. Information on the distribution of coal according to the overburden ratio exists for the Powder River Basin area of Montana and for Illinois; see Matson and Blumer [8] and Simon and Smith [14]. The data are plotted on log-probability paper and are shown in figures 2.6 and 2.7. Again the fit is good over most of the range but diverges in the tail. Since the tails represent the highest-cost coal, which in the western and midwestern regions will not be used for hundreds of years, the lognormal approximation appears adequate. The Montana distribution is assumed to hold for the western states, and the Illinois distribution for the midwestern states.[18]

The variances are estimated from the data in figures 2.6 and 2.7. The

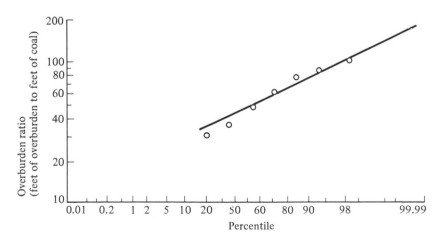

Figure 2.6 Distribution of strip-mineable coal according to overburden ratio in Illinois (Source: J. A. Simon and W. H. Smith, "An Evaluation of Illinois Coal Reserves Estimates," *Proceedings of the Illinois Mining Institute,* 1968)

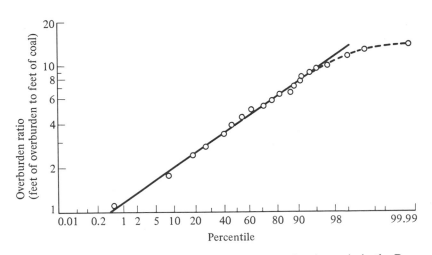

Figure 2.7 The distribution of coal according to overburden ratio in the Power River Basin (Source: Robert E. Matson and John W. Blumer, *Quality and Reserves of Strippable Coal, Selected Deposits, Southeastern Montana,* Montana Bureau of Mines and Geology, Bulletin 91, December 1973)

means are calculated separately for each state. The Bureau of Mines has estimated how much strip coal lies at overburden ratios of less than what they call the economic limit and how much lies at greater ratios. This economic limit varies from state to state. These data are utilized to estimate the mean in the same way that the mean of the seam thickness distribution was obtained; see U.S. Bureau of Mines [18].

In the east a more elaborate procedure was used. There the typical strip-mining operation is on a hillside. Overburden ratios increase as mining proceeds deeper into the hillside. The U.S. Geological Survey estimates mineable strip reserves by calculating the slope of the hill and the thickness of the coal seam. Given the slope and seam thickness, they can calculate how much coal is available at various overburden ratios.[19] Utilizing USGS data and procedures, we were able to approximate the distribution of coal according to overburden ratio for the eastern coal-producing states. The actual distribution is described in appendix A.

The Distribution of Tons According to the Cost of Mining

We now have all the information necessary to construct the distribution of coal according to the cost of mining. The distribution for deep coal according to the cost of mining is lognormal. The mean of the log of cost is $\log K_D - 1.1071 \log Th$, and the variance is $(1.1071)^2 \, \sigma^2_{\log Th} + \sigma^2_{\log \epsilon}$.

For strip coal R and η are distributed lognormally, so log $(AC - 0.96)$ is also distributed lognormally with mean log $K_S + 1.63$ log R and variance $\sigma^2_{\log R} + (1.63317)^2\sigma^2_{\log \eta}$. The K_D and K_S represent the estimated contants.

There is one remaining complication, again due to depletion. The best coal has already been mined out. Therefore the remaining coal must be at least as expensive as today's level of cost. The distribution of coal remaining in the ground according to the cost of mining is therefore a truncated lognormal. This is illustrated in figure 2.8.

The vertical axis measures tons available, and the horizontal axis measures cost. The area under the curve between any two cost levels is the amount of coal available in that cost interval.

If C_1 were the cost of mining today, then after mining T tons, the cost would be C_2. Costs of alternative levels of output can be estimated in this way. The coal represented by the area under the curve to the left of C_1 has been eliminated through previous mining.

The analytic expression for the distribution is

$$\frac{\phi(\log c)}{1 - \int_{-\infty}^{\log \bar{c}} \phi(\log c)\, d\log c,} \tag{2.11}$$

where ϕ is the normal frequency function and the denominator reflects the truncated nature of the distribution.

The estimation process described is repeated for each state and for each sulfur category. For each state there are twenty cumulative cost curves: a strip and a deep supply curve for each of ten sulfur categories. The state supply functions are aggregated into coal-producing regions to make integration with a demand model computationally feasible.

The Effect of Policy Changes
To simulate alternative policies, these cumulative cost curves have been made sensitive to policy variables. Factor price changes alter the cost of mining. This is equivalent to a shift in the mean of the cost distribution of figure 2.8. Strip-mining reclamation results in an added cost only to strip mining. State tax policies will increase costs and shift the cost distribution. The model has been programmed to accommodate these and other policy changes by effecting the appropriate shift in the distribution. The changes operate at the state level. The aggregate regional supply curves sum the deep- and strip-mining supply curves for each state. If strip mining is made more expensive in Colorado, for example, this will be

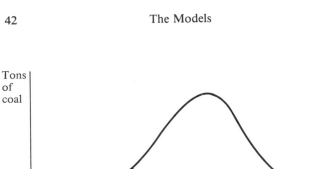

Figure 2.8 Distribution of coal according to the cost of mining

reflected by the regional supply curve. In a given cost interval Colorado will supply less coal after the tax and account for a smaller proportion of the total regional supply. Furthermore Colorado will supply relatively less strip and more deep in that interval. The state level of aggregation allows a detailed characterization of policy. The aggregation into regional supply functions assures computational ease when the coal supply model is aggregated with a demand model.

2.3 The Transport Model

The coal supply model determines price at the mine mouth. Demand, however, depends on delivered price. The missing component is transport cost, and for coal this is a significant proportion of delivered cost.

The predominant mode of transport for large-scale coal shipments is the unit train. These are shipments in railroad equipment dedicated to a particular haul between a mine and a destination. The rates charged by railroads to haul this coal result from a bargaining process between railroads and coal purchasers. Finally, the agreed-upon rate is subject to regulation by the ICC.

The ICC does not enter the process unless specific rates are challenged by one of the parties or by a third party claiming injury. A consumer group, for example, might challenge a rate. In practice a challenge to a given rate is likely when the rate charged to utility or other purchaser of coal differs substantially from rates charged to others for shipments from the same area to similar destinations. Consequently it is difficult to break an historical pattern of rates. The present pattern of rates therefore provides a good base for prediction in areas where coal shipments have been taking place for a long period of time, such as the eastern or midwestern United States. The model used to predict rates for these regions relies on rate patterns established over a period of time and in effect today.

However, the historical situation provides little guidance where shipments are just beginning. This is true today in the western United States. As discussed in chapter 1, the situation with respect to transport rates in the western United States is in flux. Chapter 3 simulates the effects of possible regulatory actions with respect to transport rates; here we describe how the base level of rates was determined.[20]

Eastern and Midwestern Rates
The first unit trains introduced in the United States were in the east and midwest. Eastern rates are best predicted by understanding the process by which these rates were set. The pattern of rates emerged in the 1960s. The dominant force was competition with natural gas. Where utilities could obtain cheap natural gas, rates charged for hauling coal were less. It was cheaper to ship a given amount of coal a given distance where the cost of natural gas available to the utility was lower. This is because the railroads were partially successful price discriminators. The rate charged today for shipments in the east or midwest reflects the pattern established then but is adjusted for increases in costs of haulage. To capture this, a model was estimated using actual unit train rates as of 1970. That model explicitly accounts for the market power of the railroads as well as factors that determine cost. To predict the 1977 rate, these rates are escalated by the American Association of Railroads Cost Index. The equation estimated was the following:

$$T = AAR \times (\alpha_0 + \alpha_1 M + \alpha_2 IMAT + \alpha_3 L + \alpha_4 J + \alpha_5(PALT - FOBCOAL) + \alpha_6 (PDUM) (PALT - FOBCOAL) + \epsilon,$$

where

T = rate per ton,
M = miles shipped one way,
$IMAT$ = annual volume shipped in tons,
L = loading plus unloading time in hours,
J = 1 if the shipment is by more than one railroad, 0 otherwise,
$PALT$ = price of natural gas available in 1970 in ¢/million Btu,
$PDUM$ = 1 if eastern shipment, 0 otherwise,
$FOBCOAL$ = mine mouth price of coal in 1970 in ¢/million Btu,

ARR = percentage increase in the American Association of Railroad Index,

ϵ = disturbance term.

The variables M, T, L, and J reflect cost-determining variables. The variables $PALT$-$FOBCOAL$ are the total rent appropriable by the railroad since it is the total amount a utility would pay to haul coal before it chose to build gas-burning capacity. The parameters α_5, α_6 measures what proportion of the rent, on average, the railroads obtained. This parameter is estimated separately for the east and for the midwest since barge competition in the midwest led to more competitive rates than in the east. The resulting equation is

T = [0.1331 + 8.6915$IMAT$ + 0.004595M + 0.009543L
(*t*-statistic)   (0.6294)   (0.1227)   (9.124)   (3.2339)

+ 0.0188($PALT$ − $FOBCOAL$)
(1.7909)

+ 0.0358$PDUM$($PALT$ − $FOBCAL$)   (2.12)
(3.9933)

+ 54J]AAR.
(0.6136)

It is clear from equation (2.12) that railroads were able to discriminate. The value of α_5 plus α_6 indicates that railroads in the east were able to capture, on average, 22 percent of the appropriable rent.[21] However, in 1970 the average rent ($PALT$ − $FOBCOAL$ − T) was small. The structure of rates as of 1975 thus includes a base level of rents that was small.

Western Rates

In the western United States three classes of rates must be considered. First, there are rates established before 1970; they have simply been escalated to reflect cost increases. Then there are rates established between 1970 and 1975; these are, on average, higher than the earlier rates. This reflects the improved bargaining circumstances of railroads in the period when fuels competing with coal were either unavailable (gas) or available at a much higher cost (oil, nuclear). Finally, rates being published today are even higher than the 1970 to 1975 rates. This reflects both improved bargaining circumstances and costs incurred in expanding capacity; see Zimmerman [25].

In face of the uncertainty surrounding future rates, we adopt the following strategy. We estimate rates according to their historical behavior. These rates serve as the base level of transport costs. The rates then reflect higher rents east of the Mississippi River. The new patterns had not yet emerged in Texas and other locations. These new patterns are reflected in subsequent analysis by simply raising the level of western rates.

The transport cost equations for western coal are as follows:

$$T = [-2.3846 + 17.0569IMAT + 0.008889M$$

(t-statistic)   (-2.7090)   (0.3437)   (9.1556)

$$+0.03657L + 2.3789DC]INFL. \tag{2.13}$$

(2.7950)   (5.4549)

where

DC = variable with value = 1 if shipment moves east of the Mississippi River,

$INFL$ = inflation index = 80 percent of AAR index (this formula applied by Burlington Northern Railroad to western rates).

Equations (2.12) and (2.13) are used to generate a transport cost matrix used as input to the linear program. This matrix is appendix table D.1.

2.4 The Coal Demand Model

The supply model together with the transport rate equation yields a delivered cost of coal. The missing elements are demand and a market-clearing mechanism. To capture demand, we have adapted a well-known model for our purposes.

Sources of Demand

We have seen that the electric utility sector accounted for the bulk of coal consumption in the United States. Steam coal is also used in the industrial sector. The steel industry consumes large amounts of coal in coke production as well as steam coal. Coking coal, when heated in the absence of oxygen, cakes or forms coke. This metallurgical coal must be low in ash and sulfur. Furthermore the volatility content strongly influences the value of the coal. Low-volatile, low-sulfur, low-ash coal earns a substantial premium and thus is not used for steam raising where only heat and sulfur content matter.[22] The bulk of coal exports are metallurgical coal. After disappearing in the 1960s, steam coal exports to Europe began to reappear only recently.

The Demand for Electricity

The demand for coal is derived from the demand for electricity and from the demand for industrial products. The model we have adapted for the purpose of generating coal demand deals with the derived nature of coal demand. The model used is the Regional Electricity Model (REM) developed at MIT by Martin Baughman, Paul Joskow, and Dilip Kamat (The MIT Press, 1979).

The demand model begins by estimating the demand for energy. In the residential and commercial sectors this is done through an econometric equation that expresses the demand for energy per capita as a function of a weighted average of energy prices, income, and population. This equation is estimated from combined time-series and cross-sectional state data. A lagged adjustment structure is estimated to separate long-run and short-run effects. These demands for energy are then broken by fuel-split equations into demand for fuel types. The proportion of total energy demand satisfied by a particular fuel is related to the relative prices of each type of fuel.

The industrial sector is treated in a similar way. Total energy demand is first estimated as a function of an energy price index and value added in manufacturing. This total demand is allocated to states by a cross-sectionally estimated location equation. This equation estimates the relative share of each state in total energy consumption as a function of prices, income, and population relative to a reference state. Finally, the demand per state is divided into fuel types by fuel-split equations.

The output of this procedure is a demand for coal in the industrial sector and the demand for electricity. The former is a demand for direct use of coal. The latter must be further decomposed into the demand for the different fuels used to generate electricity.

The Demand for Coal in the Electric Utility Sector

The essence of the electricity model is the economics of plant expansion first worked out by Turvey [16]. The cost minimization problem is complicated in the electric utility sector by the pattern of electricity demand. The demand level fluctuates throughout the day, as well as from month to month. For example, the highest level of demand typically occurs during the hottest day of the year, when air conditioning is in high use. The problem from the standpoint of the utility is that it must meet such high levels of demand only a few days a year.

The solution to this problem is to use different kinds of plants. Clearly a utility does not want to build a capital-intensive plant and then use it only five days a year. To satisfy those peak demands, it will build a less capital-intensive plant that uses fuel more intensively. Then the utility will pay for fuel only on those days when the plant is in use, thus carrying smaller capital costs. On the other hand, there is some amount of electricity that they must generate throughout the year, day in and day out. To satisfy this base-load demand, utilities will choose the more capital-intensive and less fuel-intensive plants. Because these base-load plants realize high rates of utilization, capital costs per unit of output are reduced. Since the plants operate for much of the year, the variable costs, particularly fuel, are lower than for peaking units.

This decision about plant type is illustrated graphically in figures 2.9 and 2.10. Figure 2.9 presents a load-duration curve. That curve describes how much power is demanded for what fraction of the year. For example, point H_2 indicates that for H_2 hours in the year the demand is at least equal to X_1 megawatts.

Figure 2.10 summarizes the plant economics. Each plant can be characterized by a capital cost and variable cost. The latter consists of fuel charges and maintenance and operating costs. The peaking plant curve, for example, represents a plant with low capital costs and relatively high fuel costs. Therefore average cost declines slowly. The base-load curve is the opposite. It has a high capital cost, and, as utilization rates increase,

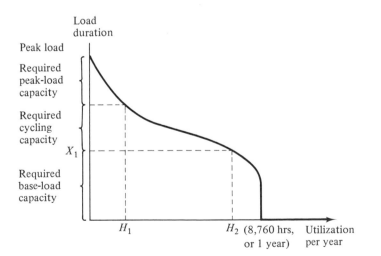

Figure 2.9 Idealized load-duration curve (Source: M. L. Baughman, P. L. Joskow, and D. P. Kamat, *Electric Power in the United States: Models and Policy Analysis,* Cambridge, Mass.: The MIT Press, 1979)

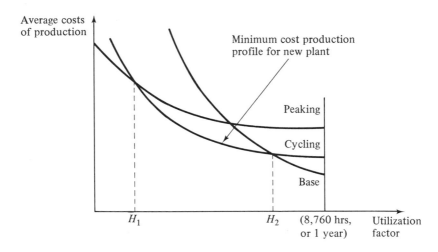

Figure 2.10 The economics of plant choice (Source: M. L. Baughman, P. L. Joskow, and D. Kamat, *Electric Power in the United States: Models and Policy Analysis,* Cambridge, Mass.: The MIT Press, 1979)

its average cost declines rapidly. Turvey has shown that the demand of H_1 hours or less is most economically served by peak plants, and similarly for the others.

This describes the optimum case. If capacity already exists, then variable cost is all that matters. Therefore existing plants will be used until variable costs become so high as to preclude their use.

Fuel costs and the expected costs of the various plants are forecast in the model for the relevant planning period. The expectations are generated by smoothing trends and projecting the trends over the planning period. The model thus uses a simulation approach in attempting to capture actual behavior of utilities, rather than assuming the utilities are solving a dynamic programming problem with perfect foresight.

The Baughman-Joskow-Kamat model uses this expansion algorithm to determine the optimum capacity if there were no existing plants. It actually builds only the difference between the optimum and existing plants, allowing for the retirement of existing but obsolete plants.

The demand model assumes utilities use a trend-adjusted moving average to forecast demand levels. The utility uses these forecasts to determine how much capacity it needs. In any year actual demand is determined by the econometric model just described. The demand forecasts are adjusted over time as actual demand evolves. Again, the simulation approach attempts to capture actual utility behavior.

In any given year with fixed plant capacity the problem is to minimize the cost of meeting the actual demand. This is done by using most frequently those plants that cost the least. In any given year this minimization routine determines actual plant use. Each plant type is ranked in terms of fuel and operating costs. The total demand is met by going down the list of plants from low to high cost until the total demand is met.

As we have modified the model, the plant types available to the decision maker are the following: coal plants without stack gas scrubbers, coal plants with scrubbers, nuclear plants, oil and gas plants, and hydroelectric plants. The expected costs of generation for each of these plants are compared when making capacity expansion decisions. The decision as to type of coal plant depends upon the premium paid for low-sulfur coal. The electric utility model receives from the coal model two coal prices: one price for coal that satisfies pollution regulations without scrubbers and one price for high-sulfur coal that requires scrubbers to meet pollution requirements. The expected costs of plants with scrubbers are compared to plants without scrubbers over time, and the least-cost alternative is chosen.

We have added an additional decision to the model—the choice of whether to retrofit existing coal plants. In each period the calculation is made whether it would be cheaper to meet sulfur regulations by burning low-sulfur coal for the life of the plant or by adding a scrubber to an already existing coal-burning plant.[23] The calculation compares the present value of the expected additional costs of burning low-sulfur coal with the costs of retrofitting an existing plant.

The Regulatory Section of the Demand Model
The price of electricity that is input to the demand equations in the model is determined in a simple regulatory and financial submodel. This price results from applying an allowed rate of return on the rate base to the total capital invested. It is important to bear in mind that this is a regulated price and not marginal cost. Policies that raise costs in the long run can lower prices in the short run because of the way in which regulated prices are determined. The allowed rate of return is calculated as a weighted average of the cost of capital. The values of the various financial parameters are presented in table A.2. The rate base includes the cost of plant and equipment used in generating capacity, as well as transmission and distribution equipment. Whereas the expansion decision is based on expected costs, the rate base measures actual costs. The cost of a nuclear unit entered into the rate base in a given year, for example, is a weighted

average of realized costs over the previous ten years. The weights are the percentage of total project cost expended in each year of a typical nuclear construction project.

Two prices of electricity are fed to the demand model. One is the residential-commercial price; the other is an industrial price. The ratio of these two prices is equal to the ratio of cost of supplying electricity to these two demand categories. The difference in cost is due to the difference in transmission and distribution cost.

Transmission and Distribution

Transmission and distribution (T & D) requirements and operation and maintenance for the T & D sector are related econometrically to the characteristics of the service such as land area, type of service (residential, light, and power), number of customers, and total electric load and load density.

Metallurgical and Export Coal

The final category of coal demand is in steel making. This includes both exported and domestically used metallurgical coal. The bulk of coal exports are metallurgical coal, although over the next 10 to 20 years this could change substantially as steam coal exports assume a larger role. These metallurgical demands are relatively price inelastic. The demand for coal used in steel-making depends much more on the demand for steel than on the price of high-quality coal. We treat these demands as exogenous. We specify a total demand for each of these categories. This demand is then broken out by sulfur content and assigned to supply regions on the basis of historical patterns, which are assumed to be fixed over time.

2.5 The Link between the Supply and Demand Models

The first step in the link between the supply and the demand models is a linearization of the cost curves. The highly nonlinear cumulative cost functions are approximated by step functions. A typical approximation is shown in figure 2.11. The program chooses the level of output from each step. Clearly lower-cost steps are used before higher-cost steps.

The linearized supply curves and the demand model are linked together through a linear programming formulation. The supply and transport models yield the cost of producing and transporting coal from supply region to demand region. The demand model yields coal demand by the

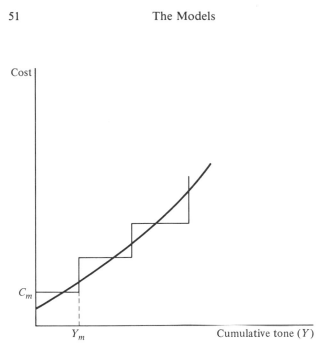

Figure 2.11 Linearized cost curve

nine U.S. census regions. This is then disaggregated into twelve smaller demand regions and fed to the linear program as constraints. The linear program minimizes the cost of meeting these demands, given mining cost, transport cost, and subject to additional constraints on the sulfur content of the coal.

As explained earlier, demand in year t is a function of prices in year $t - 1$. That is, demand adjusts with a one-year lag. Given the nature of purchasing decision by electric utilities, this is not unrealistic. Furthermore price changes from year to year are relatively small in this long-run model, and demand elasticities in the short run are small, so we are really never far from the equilibrium set of prices and outputs.

The linear program is solved year by year. However, mining costs in each year are updated, given the additional cumulative output represented by last year's output. Thus the costs in any year are a function of the sum of previous years' outputs. Each year a new set of step functions is calculated. In this way depletion is explicitly accounted for.

This depletion phenomenon partially captures the interactions through time. Further the demand model bases its decision on expected future costs. However, the solution to the linear program is not a full dynamic solution. Several reasons militate against a full-scale dynamic version of this model. First, the computational cost is extremely high. Since our goal was to analyze many different policies, a full dynamic solution, given the detail on the policy side, is prohibitively costly. Second, we examine *ex post* how wrong the static solution is likely to be. The bias turns out to be small, and adjusting for the bias simply reinforces our conclusion. Lastly, the full dynamic version is not in keeping with the simulation nature of the Baughman-Joskow-Kamat model. In essence it would assume complete knowledge about the future on the supply side with simulated rules-of-thumb on the demand side.[24]

The Program

$$\min \sum_{i=1}^{6} \sum_{j=1}^{12} t_{ij} \sum_{s=1}^{8} \sum_{k=0}^{1} Z_{ijsk} + \sum_{i=1}^{6} \sum_{s=1}^{8} \sum_{m=1}^{6} C_{ism} Y_{ism}. \tag{2.14}$$

$$\text{s.t.} \sum_{m=1}^{6} Y_{ism} - \sum_{j=1}^{12} \sum_{k=0}^{1} Z_{ijsk} = 0, \quad \text{for } i = 1 \ldots 6, s = 1 \ldots 8; \tag{2.15a}$$

$$\sum_{i=1}^{6} B_i \cdot \sum_{s=1}^{8} Z_{ijsk} \geq D_{jk}, \quad \text{for } j = 1 \ldots 12, k = 0, 1; \tag{2.15b}$$

$$\frac{2{,}000 \cdot 2 \cdot \sum_{i=1}^{6} \sum_{s=1}^{8} P_s Z_{ijso}}{\sum_{i=1}^{6} B_i \cdot \sum_{s=1}^{8} Z_{ijso}} \leq S_j, \quad \text{for } j = 1 \ldots 12, \tag{2.15c}$$

$$Y_{ism} \leq \overline{U}_{ism}, \quad \text{for } i = 1 \ldots 6, s = 1 \ldots 8, m = 1 \ldots 6, \tag{2.15d}$$

$$Z_{ijsk}, Y_{ism} \geq 0, \quad \text{for } i = 1 \ldots 6, j = 1 \ldots 12, s = 1 \ldots 8, k = 0, 1,$$
$$m = 1 \ldots 6.$$

where

t_{ij} = transport cost in dollars per ton from supply region i to demand region j,

Z_{ijsk} = shipment of coal in tons from supply region i to demand region j of sulfur type s and scrubbing activity k,

Y_{ism} = production in region i of coal of sulfur category s from step m,

C_{ism} = cost of coal (i, s) from step m,

B_i = Btu value per ton of coal in region i,

D_{jk} = demand in region j for coal from scrubbing category k, in Btu,

P_s = sulfur content in percent by weight of coal of sulfur category s,

$\overline{U}_{ism}$= length, in tons, of step m in region i, sulfur category s.

The Objective Function

The objective function simply minimizes the cost of mining and transporting coal. The Z_{ijsk} variables are shipments from supply to demand regions. These are summed over all sulfur and scrubbing categories. The t_{ij} is the cost of transport estimated in the transportation submodel. The Y_{ism} are production levels on the steps of the linearized cost functions. The C_{ism} is the cost for that step estimated in the cost model.[25]

The Constraints

The first set of constraints and the last two sets of constraints are "housekeeping." The first set forces quantities produced (Y_{ism}) to equal quantities shipped (Z_{ijsk}).

The next-to-last set of constraints simply reflects the stepwise approximation to the cost curves. The $\overline{U}_{ism}$ are the quantities of coal in each step, and production from each step must be less than that level. The last set of constraints is the set of non-negativity constraints.

The Demand Constraint

The second set of constraints forces the model to satisfy the demand in each region as determined in the demand model. The demand model calculates two sets of demand in each region as denoted by the subscript k. The first set is coal demand in plants using scrubbing devices. The second set is for coal demand in plants not using scrubbers.

The decision whether or not to build a scrubber is made in the demand model. The utility model decides whether it would be cheaper to satisfy sulfur pollution regulations with a scrubber or with low-sulfur coal. Once having decided to build a scrubber, it always uses it. However, if it does not build a scrubber, it can at a cost penalty retrofit the plant with a scrubber.[26]

All decisions in this long-run equilibrium model are based on current long-run marginal costs. This is appropriate when considering long-run policy issues. However, in the past several years changes in the industry have been rapid, and the difference between marginal and average cost is great. Many mines were opened when factor prices and transport costs were very different. These mines, once opened, disregard past capital costs in their decisions. They do not shut down even if they would not be opened today. To reflect this decision, we calculate a set of market sensitive demands. These are the demands sent to the linear program as constraints. Total annual demand is calculated by region in the demand model. However, subtracted from that demand is what was fixed before the model runs began, before 1975. We assume that the coal shipments fixed by 1975 decline at a rate of 5 percent per year, the historical rate of mine closures. We take the total amount fixed by 1975 as a fixed fraction of 1975 demand.[27] In each period we optimize the production and distribution pattern for incremental demand equal to the net increase in demand over the fixed 1975 levels, plus the cumulative amount of mine closures through that year. We do not restrict the program in subsequent years. The model does have the option of assuming that shipments resulting from the optimization routine in years after 1975 are fixed by contracts for a given number of years. However, we have opted for the simplest form.[28] Incremental demand may be written as

$$\begin{aligned} ID &= D(t) - \delta D(0) + 0.05[\delta D(0)]t \\ &= D(t) + \delta D(0)[0.05t - 1], \quad \text{if } t < 20, \\ &= D(t), \quad \text{if } \geq 20, \end{aligned}$$

where

$\quad ID =$ incremental demand,
$\quad \delta =$ fixed fraction of 1975 demand, taken as 0.9,
$\quad D(0) =$ 1975 demands,
$\quad D(t) =$ demand in year t.

The Sulfur Constraints
The coal used must satisfy restrictions on sulfur emissions. These constraints reflect the impact of federal and state air pollution regulations. Actual regulations are extremely complex and are applied on a plant-by-plant basis. It would be computationally infeasible to deal with the complexity of these regulations. The situation is made even more difficult since there is a wide gap between the passage and enforcement of the law. We have adopted a computationally simple approximation that captures the essence of the regulations.

As we indicated in chapter 1, there are four sets of applicable air pollution regulations. New Source Performance Standards limit the sulfur content of coal used in plants coming on stream after 1975. These plants have to emit less than 1.2 lb of sulfur dioxide per million Btu burned. The utilities could meet that standard by burning coal low in sulfur or by using stack gas scrubbing to remove the sulfur from the effluent gas. Furthermore revised regulations force scrubbing in plants under construction after 1978. The third category of constraints represents the State Implementation Plans (SIP). In many states the SIPs limit emissions on a plant-by-plant basis to meet general ambient air standards.

These standards were supposed to have gone into effect on June 30, 1975, at the latest. However, in many areas of the country these plans are not being enforced. Current schedules call for complete enforcement by 1983. In most areas NSPS standards are more stringent than SIP standards and are therefore binding on new plants. Lastly in areas that are already below the applicable standards, there are nondegredation requirements: plants cannot further increase air pollution in the area. Again, standards can be met by scrubbing or by burning any mixture of sulfur contents that satisfy the standards.

The sulfur constraints of the model state that the emissions from incremental coal, expressed as lb SO_2 per million Btu burned, must be less than an appropriate standard. In light of a complicated set of standards, the issue is how to choose the appropriate value for the constraint $\overline{S}_j$. In essence we calculate outside the linear program what the constraint is for each category of sulfur regulation. The tightest constraint is the binding constraint, which is entered as $\overline{S}_j$ on the right-hand side of the sulfur constraint.

Current levels of actual emissions, in terms of lb SO_2 per million Btu burned, were calculated for each region. This was the base-level standard; emissions from incremental coal must be less than this average. In this way we assure compliance with nondegredation standards. Second, we assume all incremental coal must meet a standard of no more than 1.2 lb SO_2 by 1985. Between 1975 and 1985 the standard begins at current levels and declines linearly to the 1.2 lb standard in 1985. In areas where emissions are already below 1.2 lb, the actual emission level becomes binding. Finally, BACT requirements are imposed by forcing scrubbing for plants scheduled to come on-stream after 1983.[29]

The reason for the phased tightening in the sulfur standards is the following: in reality not all incremental coal must meet NSPS. This is because incremental coal is not all being used in plants coming on-stream after 1975. Some of the new contracts will be for coal used in plants built before 1975. This means we overestimate the requirements for NSPS. This problem diminishes over time as new plants replace old ones. To ease that problem, we have implemented the NSPS slowly, becoming fully effective in 1985.[30]

In each year, in each region, two coal prices are transmitted back to the demand model. The average price of coal used in plants with scrubbers corresponds to the cost of high-sulfur coal. Low-sulfur coal prices are taken as the average cost of coal in plants without scrubbers. These prices are fed back to the utility model to determine plant choices. The average price is the weighted average coal price of all the coal consumed in that category of demand. It represents the minimum cost combination of coals that meet the sulfur constraint. The average price of that combination represents a weighted average of incremental prices for each individual coal type. The average price is what it would cost, at the margin, to consume a mixture of coal that satisfies pollution regulation.

The disaggregation into two demand categories allows us to interact with the demand model in an efficient way. Rather than giving the demand model all possible coal prices, only two prices are passed back.

Exogenous, Endogenous, and Policy Variables

The cost of mining coal will change over time as factor price changes interact with depletion. In all the policy analyses reported in following chapters we specify an exogenous rate of change in capital costs and in wages.

The demand model endogenously determines the price of electricity, the demand for electricity, and the demand for coal. The demand model is driven by exogenous inputs on fuel prices other than coal, capital costs, and operation and maintenance costs, as well as a host of other technical input data.

In this model policy is simulated by changing the values of exogenous inputs. We assess the impact of higher transport costs by raising rates predicted by the transport model; sulfur pollution laws are examined by changing constraints in the linear program; taxes are examined by altering the level of state taxes; and so forth.

Appendix B presents the exogenous data and the initial set of policy parameters. In later chapters, when we analyze the effects of particular policies, we will test the sensitivity of model results to the exogenous inputs.

References

1. American Institute of Mining Engineers. *Mining Engineering Handbook,* Summer 1973.

2. Baughman, M. L., P. L. Joskow, and D. P. Kamat. *Electric Power in the United States: Models and Policy Analysis.* Cambridge, Mass.: The MIT Press, 1979.

3. Data Resources, Inc. "New Source Performance Standards. The EPA Considers Its Options" (by S. E. Martin). *Coal Review.* (May 1979): 17–27

4. Gordon, R. L. *U.S. Coal and the Electric Power Industry.* Baltimore: The Johns Hopkins University Press, 1975.

5. Hall, R. E., and D. W. Jorgensen. "Tax Policy and Investment Behavior." *American Economic Review,* 57 (June 1967): 319–414.

6. Hausman, J., and D. Wise. "Social Experimentation, Truncated Distributions and Efficient Estimation." *Econometrica,* 45 (May 1977): 919–938.

7. Hotelling, H. "The Economics of Exhaustible Resources." *Journal of Political Economy,* 39 (April 1931): 137–175.

8. Matson, R. E., and J. W. Blumer. *Quality and Reserves of Strippable Coal, Selected Deposits, Southeastern Montana,* Montana Bureau of Mines and Geology, Bulletin 91, December 1973.

9. McGraw-Hill. *Keystone Coal Manual.* New York: McGraw-Hill, various years.

10. Nielson, G. L. "Coal Mine Development Expansion Survey." *Coal Age,* 82 (February 1977): 83–92.

11. NUS Corporation. *Coal-Mining Cost Models, Volume I, Underground Mines.* EPRI final report no. EA-437, February 1977.

12. Paul Weis Company. *Economic Study of Coal Reserves in Pike County, Kentucky, and Belleville District, Illinois,* January 1972.

13. Rumfelt, H. "Computer Method for Estimating Proper Machinery Mass for Stripping Overburden." *Mining Engineering,* 13 (May 1961): 480–487.

14. Simon, J.A., and W. H. Smith. "An Evaluation of Illinois Coal Reserve Estimates." *Proceedings of the Illinois Mining Institute, 1968.*

15. Stefanko, Robert, Ramani, R. V., and Ferko, Michael R. *An Analysis of Strip Mining Methods and Equipment Selection. Research and development report no. 61,* Office of Coal Research, May 29, 1973.

16. Turvey, R. *Optimal Pricing and Investment in Electricity Supply.* Cambridge, Mass.: The MIT Press, 1968.

17. U. S., Bureau of Mines. *Cost Analyses of Model Mines for Strip Mining in the United States.* Information circular 2535. Washington, D.C.: Government Printing Office, 1972.

18. U. S., Bureau of Mines. *Strippable Reserves of Bituminous Coal and Lignite in the United States.* Information circular 8531. Washington, D.C.: Government Printing Office, 1971.

19. U. S., Bureau of Mines. *Basic Estimated Capital Investment and Operating Costs for Underground Bituminous Coal Mines.* Information circulars 8632 and 8641. Washington, D.C.: Government Printing Office, 1974.

20. U. S., Bureau of Mines. *The Reserve Base of U.S. Coals by Sulfur Content.* Information circular 8680. Washington, D.C.: Government Printing Office, 1975.

21. U. S., Senate, Committee on Interior and Insular Affairs. *Coal Surface Mining and Reclamation.* Washington, D.C.: Government Printing Office, 1973.

22. White, D. E. *The Application of Mathematical Programming Techniques to the Integration of a Large-Scale Coal Energy Model.* Ph.D. dissertation. Department of Civil Engineering, Water Resources Program, MIT, 1980.

23. Williams, C. H., Jr. "Planning, Financing, and Installing a New Deep Mine in the Beckley Coal Bed." *Mining Congress Journal,* 60 (August 1974): 42-47.

24. Zimmerman, M. B. "Modeling Depletion in a Mineral Industry: The Case of Coal." *Bell Journal of Economics,* 8 (Spring 1977).

25. Zimmerman, M. B. "Rent and Regulation in Unit-Train Rate Determination." *Bell Journal of Economics,* 10 (Spring 1979).

26. Zimmerman, M. B. "Estimating a Coal Policy Model." *Advances in the Economics of Energy and Natural Resources,* vol. 2. Greenwich, Conn.: JAI Press, 1979.

3   Basic Forces

The development of the coal industry will be affected by a variety of forces. Some of these forces are central to the coal industry itself. Mining wages, sulfur regulations, and coal transport rates are among those endogenous to the coal sector. Other forces that seriously will affect the industry reflect developments outside the coal sector. The future of nuclear power and oil price increases, for example, will determine the demand for coal. In this chapter we focus on the internal supply influences shaping coal costs and supply. The goal is to understand the forces shaping production and distribution patterns within the industry.

The analysis asks a series of hypothetical questions. To isolate the effects of environmental policies and economic factors, we peel away policies and factor price changes and simulate the model as if those policies had not been enacted or the changes had not taken place. We ask, for example, what would the coal industry look like if there were no Clean Air Act? We then implement the policy and examine the net change. In this way we can measure its quantitative impact on output and prices. To examine the effects of factor price changes, we arbitrarily assume different relative prices and simulate the model. By comparing results in different cases, we can isolate the impact of economic forces on production and distribution of coal. Our goal is not to predict but rather to gain insight into the forces shaping the industry.

Before examing the issues, a cautionary note is in order. The supply model is an equilibrium model. It assumes that coal is traded at long-term contract prices and does not address year-to-year fluctuations caused by unanticipated developments. There is no way to forecast deviations from expectations.

We have not considered short-run bottlenecks because this would necessitate a far more detailed study and a far larger budget than are at our disposal. Rather, here we ask what long-run equilibrium would look like. In that way we examine the pressures that arise as a particular policy is implemented. In several cases the pressures are extreme. For example, levels of production that would completely transform regions of the United States are called for. It is not likely that these could be achieved in the period of time considered. In such instances the cases underscore the extreme difficulty and disruption a particular policy course would entail; they do not depict the equilibrium. Long-run equilibrium models are useful for highlighting the pressures in a system, not for predicting outcomes in any particular year.

There is another important aspect of the equilibrium nature of the model. By allowing the model to solve for long-run equilibrium prices, we obtain in essence a lower limit on the price effects of any particular policy. Since we allow the coal industry to adjust optimally, we obtain an estimate of the price of electricity under the most favorable circumstances. The bias is on the down side; a price reflecting short-run bottlenecks would be higher.

We maintain this conservative bias throughout this chapter in another important way. We are attempting to analyze pressures, costs, and structural changes incurred by coal policy actions. We want to bias the results in such a way that we can be confident about our lower-bound estimates. We want to assume a set of circumstances that is likely to minimize the costs of adjustment. In these scenarios we are thus purposely optimistic about nuclear power in the long run. The current difficulties faced by nuclear power are severe. Public acceptance is at a low ebb. Operating rates have been far below expectations due to frequent shutdowns necessitated by safety problems and regulations. To reflect these current difficulties, we constrain nuclear power in the near term. We assume first that until 1988 only those plants already announced can be completed. This is quite realistic and reflects the ten-year construction lag. We further assume that no new announcements are made until 1983. In effect no nuclear power plants not already announced can be on-stream before 1993. In light of recent developments this is a likely scenario.

This pessimism about the short run means that for the next thirteen years the path of nuclear power is completely determined. However, there are many reasons why this could change in the longer run. Public attitudes toward nuclear power could change. Operating rates could improve. Original expectations were for 80 percent operating rates; now they are at 65 percent. It is impossible to predict how attitudes and operating rates will evolve over the next twenty years. Thus we adopt a strategy of bounding the analysis. In this chapter and the next we are long-run optimists.

We assume availability rates improve to 73 percent.[1] We allow the model to choose the amount of nuclear power generation after 1993, subject only to the expected costs and constraints shown in appendix B. In chapter 5 we examine the worst case as well as the sensitivity of the results to key assumptions about capital costs and operating rates. In sum nuclear capacity over the next thirteen years is largely determined. After that, significant changes could occur, and we examine the range of possible outcomes.

The bulk of our computer runs were conducted using oil price forecasts consistent with [6]. Already events in the world oil market have overtaken these assumptions. Prices will be higher than those assumed here. This again only serves to reinforce the character of our analyses and further buttresses our conclusions.

In sum, we view the model as a way to examine the logical consequences of a particular action. It presents a set of outcomes that must be tolerated if a particular policy is implemented. In this way the trade-offs involved in retarding nuclear power, in slowing strip mining, and so on, can be quantified. Because the model assumes full adjustment in long-run equilibrium, and because we are optimistic about alternatives, our estimates of costs must be seen as lower bounds. Short-run difficulties and less favorable alternatives could raise costs substantially above the levels indicated here.

3.1 Economic Factors in the Industry

Since the early 1970s it has been apparent that western coal would play a large role in satisfying the U.S. demand. Tables 3.1 and 3.2 summarize reasons for the attractiveness of western coal. These tables present the estimated rate of cost increase over time in the various supply regions for different quality coals, if output were to remain constant at 1973 levels. Thus, if output rates of northern Appalachian high-sulfur deep coal were to remain constant, it would take twenty years to realize a cost increase of 10 percent.[2]

The tables describe the underlying supply curves of chapter 2. If output rates were to double, then costs would increase by 10 percent in northern Appalachia in ten years. Now, if at the same time the output rate is doubling low-sulfur is substituted for high-sulfur coal, we would experience a high increase in eastern low-sulfur coal costs. This would cause a movement to western coal, particularly strip-mined coal for which cost in-

Table 3.1 Estimated percent increases in cost over time at current rates of output for eastern deep coal

	5 Years	10 Years	20 Years	30 Years	50 Years
Northern Appalachia					
Medium sulfur	4.5	8.6	17.0	25.0	41.0
High sulfur	2.3	5.2	10.0	15.0	24.9
Southern Appalachia					
Low sulfur	7.9	15.0	31.8	51.2	98.6
Medium sulfur	6.9	13.2	25.7	40.9	68.0
High sulfur	2.9	5.9	11.4	16.8	29.6
Midwest					
High sulfur	0.6	1.5	2.7	4.1	6.9

Table 3.2 Estimated percent increases in cost over time for midwestern and western surface-mined coal at current rates of output

	5 Years	10 Years	20 Years
Midwest			
High sulfur	10.1	15.6	25.7
Montana-Wyoming[a]			
Low sulfur	1.3	2.5	8.2

[a] These figures represent cost increases at five times the current output levels.

creases would be moderate even for very large increases in output. If sulfur regulations were not a factor, we would expect, on the basis of tables 3.1 and 3.2, to see an expansion at the margin in the production of midwestern deep mining. There costs rise gently as compared to high-sulfur coal costs in Appalachia, and underground cost increases in the midwest are far less than cost increases implied by expansion in strip mining. These tables present only the effects of depletion. They suggest broadly how the industry would develop in the absence of changes in factor prices, transport costs, pollution regulations, taxes, and so forth.

An important determinant of this evolution is labor cost. Labor productivity declined dramatically between 1969 and 1977, concurrent with increases in real wages. The net result was a large increase in the real cost of labor (see table 3.3). Because of the great labor intensity of underground mining, the rapidly rising real wage costs led to a shift to strip mining. As the supply model indicates, the supply elasticity of strip coal in the eastern United States is limited. The interaction of the response to factor prices changes and depletion thus leads to further western production.

Supply side phenomena are not the only factors acting on the industry.

Table 3.3 Trends in labor productivity and labor costs

Year	(1) Output per person-day underground	(2) Output per person-day strip	(3) Average hourly earnings in bituminus mining ($)	(4) Earnings divided by GNP implicit price deflator (1967 = 100)
1967	15.07	35.17	3.75	3.75
1968	15.40	34.24	3.86	3.69
1969	15.61	35.71	4.24	3.86
1970	13.76	35.96	4.58	3.96
1971	12.03	35.69	4.83	3.97
1972	11.91	35.95	5.32	4.20
1973	11.66	36.30	5.75	4.29
1974	11.31	33.16	6.26	4.26
1975	9.54	26.69	7.25	4.51
1976	9.10	26.40	7.78	4.60
1977	8.70	26.90	8.29	4.63
1978	8.25	25.00	9.57	4.97

Sources: Columns 1 and 2: Bureau of Mines, *Minerals Yearbook,* various years; column 3: Bureau of Labor Statistics, *Employment and Earnings, U.S. 1909-1978*; column 4: *Economic Report of the President,* January 1979.

The national pattern of coal demand is changing. The growth of the sunbelt and the western demand centers is also acting to increase the demand for western coal. These western regions would use western coal regardless of sulfur regulations simply because it is the cheapest coal available to them.

All these trends were foreseen by the industry. Eastern areas producing high-sulfur coal feared the loss of markets and consequent loss of jobs and tax revenue. The federal government began to contemplate the implications of vast development in isolated areas of the west. Articles about boom towns appeared, and anticoal-development forces were marshaled in the western coal-producing states. Taxes were raised and suits brought in an effort to hold back western coal development.[3]

In addition to local efforts to slow development, there is federal action in this direction. The Best Available Control Technology (BACT) provision of the Clean Air Act was devised to favor eastern coal. Finally, both the railroads and state legislatures saw the rising demand for western coal as an opportunity to increase revenues. Given their market power, they are raising rates to increase profits. The result is higher revenues but a slowdown in the growth of western coal production.

3.2 The Effects of Depletion and Demand Shifts

The primary change in the industry during the 1970s was the rise of production in the western coal fields. What was responsible for this growth?
The common answer is that the pressure to substitute low-sulfur for high-
sulfur coal created the demand for western production.

To see what the coal industry would have looked like in the absence of
current policy, we assume that the Clean Air Act was never passed and
that factor prices to the coal industry remained constant at 1977 levels.
We assume a sulfur standard for coal of 8 lb of SO_2 per million Btu
burned. This corresponds to 4 lb of sulfur per ton of 24 million Btu coal,
or a 5 percent sulfur standard. Almost all the coal in the reserve base is
available at this high a standard. Both average and incremental sulfur
limits were set at that level. In essence this amounts to no sulfur constraint. Furthermore we assume that BACT was never implemented; in
other words, there are no stack gas cleaning requirements. This scenario
simply takes the information contained in the supply curves of tables 3.1
and 3.2 and examines the resulting pattern of regional production. We ask
what the industry would have looked like if what occurred never happened. In that way we can assess the importance of actual events. The
resulting pattern reflects both demand shifts and the depletion over time
of key segments of the industry supply curve.

Table 3.4 shows the changes in regional output levels. The interesting
trends in the absence of sulfur restrictions are the rise of western production, the rise of midwestern output, and the decline of the Appalachian
coal fields. Montana-Wyoming, by the year 2000, accounts for almost 43
percent of national output; the midwest accounts for 28 percent; and Appalachia represents only 20 percent. To what can we ascribe these vast
structural changes?

Depletion and Demand Shifts

The Rise of the West   We can gain an understanding of the western
phenomenon by examining coal flows. Table 3.5 shows where Montana-
Wyoming coal is used. The bulk of Montana-Wyoming coal is used west
of the Mississippi in demand regions 4, 8, 9, 10, and 11. Only a small
amount of Montana-Wyoming coal is used in the midwest (region 3), accounting for 8 percent of total Montana-Wyoming production by the year
2000. Without sulfur constraints western coal is used almost entirely west
of the Mississippi River. These regions are growing rapidly and
simultaneously moving away from gas as a source of power.[4] They

Table 3.4 Regional production under the assumptions of constant factor prices and no-sulfur constraints (million tons)

Year	Northern Appalachia	Southern Appalachia	Midwest	Montana-Wyoming	Utah-Colorado	Arizona-New Mexico
1980	321.8	196.3	117.4	130.5	13.5	62.6
1985	256.3	174.7	264.8	208.5	21.8	92.3
1990	352.7	201.4	374.2	354.0	80.2	97.4
1995	275.9	171.3	569.3	556.7	54.5	124.7
2000	219.7	163.5	528.2	804.9	58.8	104.6
Actual						
1977	179.8	226.4	136.4	57.1	18.2	43.3

become large coal consumers and turn to the lowest-cost coal available: western coal. Thus a good part of actual western coal expansion is a demand-driven phenomenon.

The use of western coal in the midwest, although moderate, is an interesting development. It is quite small and reaches zero in 1995. Basically the use until 1990 reflects constrained demands, since in 1975 the midwest was consuming Montana-Wyoming coal, and, as was explained in chapter 2, we constrain an ever-declining portion of that to be shipped each year until 1990. However, by 2000 there is a midwestern demand for Montana-Wyoming coal, replacing coal previously supplied by the midwest. In the absence of sulfur restrictions, this is due solely to depletion in the midwest. Even though depletion as a whole is small in the midwest, it is sufficient by the end of the period to allow a small amount of western coal to move east of the Mississippi River.

The Rise of the Midwest   The midwest coal region, unencumbered by sulfur regulations, also substantially expands output. Again, it is instructive to examine which demand regions use midwestern coal. Table 3.6 presents that information.

Clearly the midwest serves eastern markets. Its primary markets are the midwest (3), the south Atlantic (5), and the adjoining east south central region (6). The south Atlantic historically has not been an important market for the midwest, although in the past there have been small shipments to Florida.

In the model midwestern shipments to the south Atlantic show a great deal of sensitivity and abrupt shifts. As can be seen in table 3.7, the south Atlantic market buys substantial supplies from the midwest, beginning in 1995. Until that period the bulk of their coal comes from northern Ap-

Table 3.5 Distribution pattern of tons of coal shipped from Montana-Wyoming to demand regions, selected years (million tons)

Year	Demand region					
	1	2	3	4	5	6
1980	0	0	16.6	48.7	0	0.18
1985	0	0	11.0	76.2	0	0.12
1990	0	0	5.5	138.7	0	0
1995	0	0	0	187.8	0	0
2000	0	0	64.3	230.8	0	0

Note: Excludes metallurgical and export coal. Demand region numbers correspond to those of figure 2.2.

Table 3.6 Consumption of midwestern coal under assumptions of constant factor prices and no sulfur regulations (million tons)

Year	Demand region					
	1	2	3	4	5	6
1980	0	0	55.5	12.9	9.6	31.4
1985	0	0	153.5	8.6	6.4	87.6
1990	0	0	231.8	4.3	3.2	125.4
1995	0	0	269.4	0	172.7	127.3
2000	0	0	159.7	0	251.1	107.5

Note: Excludes metallurgical and export coal.

palachia (see table 3.7). In essence the consumption of northern Appalachian high-sulfur coal drives up its price until it becomes economic to switch in the south Atlantic to midwestern coal. Due to the linear programming nature of the model, the switch is abrupt. The actual movement would be more gradual. Furthermore the shift occurs in response to a small difference in delivered costs, as discussed below.

The Decline of Appalachia   In this scenario Appalachia loses its position as the largest coal-producing region. There are two aspects to the decline of Appalachian production. The first and most striking aspect is the disappearance of southern Appalachian coal from the steam coal market. Table 3.8 presents the data of table 3.4 for the steam market alone. The steam market includes coal used by both the utility and industrial markets and excludes coal for the metallurgical and export markets. The decline of southern Appalachia is striking but not surprising. The southern Appalachian region produces low-sulfur, high-quality coal. This coal is used currently in the metallurgical market as well as the steam market. However, as table 3.1 showed, depletion is strongest in the low-sulfur seg-

7	8	9	10	11	12	Total
0	22.7	0	13.4	29.0	0	130.5
0	47.8	0	19.5	53.9	0	208.5
0	82.9	0	33.8	93.0	0	354.0
0	125.5	54.6	45.9	142.9	0	556.7
0	166.1	107.0	57.6	179.1	0	804.9

7	8	9	10	11	12
0	0	0	1.0	0	0
0	0	0	0.7	0	0
0	0	0	0.3	0	0
0	0	0	0	0	0
0	0	0	0	0	0

Table 3.7 Shipments to the south Atlantic demand region by supplying region (million tons)

Year	Supply region					
	Northern Appalachia	Southern Appalachia	Midwest	Montana-Wyoming	Utah-Colorado	Arizona-New Mexico
1980	53.9	50.0	9.6	0	0	0
1985	104.1	31.2	6.4	0	0	0
1990	155.0	59.7	3.2	0	0	0
1995	51.7	29.3	172.7	0	0	0
2000	0	6.5	251.1	0	0	0

Note: Excludes metallurgical and export coal.

Table 3.8 Regional production under the assumptions of constant factor prices and no sulfur constraints, steam market only (million tons)

Year	Northern Appalachia	Southern Appalachia	Midwest	Montana-Wyoming	Utah-Colorado	Arizona-New Mexico
1980	265.8	91.3	110.4	130.5	4.5	62.6
1985	194.3	58.7	256.8	208.5	11.8	92.3
1990	284.7	73.5	365.2	354.1	68.2	97.4
1995	199.9	29.3	569.4	556.7	42.5	124.7
2000	135.7	6.5	518.2	804.9	45.8	104.6

ment of the industry supply curve. The pressure of demand from only the metallurgical and export markets drives up cost enough to eliminate this coal from the steam market. At first southern Appalachia supplies its traditional market: the south Atlantic. But this market is then lost to northern Appalachia, and finally, as we have seen, the south Atlantic turns to the midwest.

Throughout the period the northern Appalachia producing region retains its markets in New England and the middle Atlantic states. However, these are among the smallest of the regional coal markets. Northern Appalachia declines because it loses markets in the south Atlantic, the east south central, and the midwest to production from the Illinois Basin. The closeness of this competition can be seen by comparing the delivered cost of coal to these demand regions from both the midwest and the northern Appalachian region. In 1995 the midwest becomes a big supplier to the south Atlantic region. In that year, as seen in table 3.9, the cost of coal from the midwest is equal to the cost of coal from northern Appalachia.

In summary southern Appalachia experiences an unambiguous decline in output for the steam-coal market. The northern Appalachian producing region competes closely with the midwest in the south Atlantic region. The insight these results provide is not that the midwest will gain at the expense of northern Appalachia but rather that, in the absence of sulfur regulations, these two areas would be close competitors in areas not now using their coal.

In this scenario constant factor prices and unregulated supply, together with demand shifts, were enough to cause significant structural changes within the industry. Traditional markets changed. A purely econometric model would not be capable of analyzing this type of change because the underlying structure of the industry is changing in a way that cannot be predicted by examining past behavior.

Table 3.9 Delivered cost of highest-sulfur coal to south Atlantic markets ($/million Btu)

From	Mine price	Transport cost	Total
Northern Appalachia	2.76	0.84	3.60
Midwest	2.67	0.92	3.59

Note: Delivered cost differs slightly from cost reported in table 3.10 due to the linear approximation to the true cost functions.

The changes themselves are surprising, particularly the development of the western coal fields. In the absence of factor price changes and sulfur regulations, western coal grows substantially in importance. This is due primarily to demand shifts. This scenario establishes a base level of production upon which policy actions and reactions as well as factor price changes will operate. Second, the midwest and northern Appalachia compete closely for the markets lost by southern Appalachia.

Price Changes Due to Depletion in the Absence of Sulfur Regulations
The extremely large coal output effects do not have large associated price effects. Increases in the real delivered price of coal are small due to a large degree of interregional substitution. So long as the west can expand, and the high-sulfur eastern coal can be burned, the price impacts of coal industry growth are small.

Table 3.10 presents the regional delivered price of coal in current dollars. There is only one coal price: the price of high-sulfur coal. The scenario assumes no sulfur regulations. No one is willing to pay a premium for low-sulfur coal, and this coal sells at a single regional price per million Btu.[5] The highest price increases are realized on the east coast where Appalachian coal is the main source of supply. Those regions that rely solely on Powder River Basin coal experience the smallest increases due to depletion. The real increase over the period 1980 to 1999 amounts to 21 percent in the middle Atlantic and 7.9 percent in the Pacific region; there is no discernible increase in the Powder River Basin.

Sulfur Pollution Regulations
Sulfur regulations are assumed to be the New Source Performance Standards (NSPS), equal to 1.2 lb SO_2 per million Btu. These standards are assumed to take effect gradually, so full compliance is reached by 1985. Between 1975 and 1985 the standards decline in a linear fashion from 1975 actual emissions to the 1.2 lb limit. In areas where that standard was being met in 1975, we assume that level throughout the period.[6]

Table 3.10 Delivered price of coal ($/million Btu) with constant factor prices, no sulfur regulations

Year	Demand region								
	1	2	3	4J	5	6	7	8J	9
1980	1.48	1.37	1.35	1.10	1.51	1.33	1.34	0.78	1.30
1985	1.98	1.83	1.80	1.36	2.03	1.77	1.76	1.00	1.71
1990	2.73	2.53	2.38	1.76	2.80	2.33	2.37	1.32	2.31
1995	3.80	3.54	3.18	2.32	3.88	3.14	3.33	1.82	3.07
2000	4.84	4.52	4.09	2.90	4.93	4.00	4.38	2.45	3.84
Average annual increase 1980 to 2000	6.1%	6.2	5.7	5.0	6.1	5.7	6.1	5.9	5.6
Real annual increase	0.6%	0.7	0.2	—	0.6	0.2	0.6	0.4	0.1
Cumulative real increase	12.1%	13.1	3.8	—	11.9	3.1	12.0	7.7	1.2

Note: The J indicates the census region aggregates. Region 4J in the demand model corresponds to the aggregate of demand regions 4 and 10. Region 8J consists of regions 8, 11, and 12 of figure 2.2. Inflation fixed at 5.5 percent per year for the entire period.

We first examine the effects on production. The tighter sulfur constraints accelerate the use of western coal. This is seen clearly in table 3.11, which presents regional output levels under the same constant factor price assumption of table 3.4 but with the NSPS sulfur levels. The main effects are (1) a short-term rise in the importance of southern Appalachia and a concomitant decline in the short-term importance of northern Appalachia, (2) a substantial decline in the importance of midwestern production, and (3) a rise in Montana-Wyoming output. The first result is expected since the pressure to burn low-sulfur coal causes a switch from northern Appalachian high-sulfur production to southern Appalachian low-sulfur production. This is short-lived, however, because of the run-up in cost of southern Appalachian reserves due to depletion.

The decline in midwestern output is striking. There is still competition with northern Appalachia. However, even allowing for gains relative to northern Appalachia, output in the year 2000 is 42 percent less in the Illinois Basin. Output in Montana and Wyoming is 34 percent higher. The main change in the pattern of distribution is that the midwest demand region becomes exclusively a Montana-Wyoming market. In the previous

Table 3.11 Regional production with constant factor prices and 1.2 lb/million Btu sulfur standard in 1985 (million tons)

Year	Northern Appalachia	Southern Appalachia	Midwest	Montana- Wyoming	Utah- Colorado	Arizona- New Mexico
1980	201.7	247.6	134.7	193.8	21.1	54.2
1985	161.4	317.0	79.1	349.9	32.3	81.2
1990	342.1	198.5	148.4	616.4	100.1	76.8
1995	338.5	165.8	159.6	1,009.5	52.8	49.7
2000	200.7	216.4	308.7	1,077.9	84.8	45.5
Total national production in 2000	10.4%	11.2	16.0	55.7	9.4	2.4

case the midwestern demand region drew from both the midwest and Montana-Wyoming. However, it cannot continue to do so and still meet the sulfur constraints, so it turns exclusively to western coal. This is a direct substitution for Illinois Basin coal, and production in Illinois-Indiana-West Kentucky declines. A second important effect is the substitution between western coals. Under the pressure of sulfur constraints Montana and Wyoming gain at the expense of Arizona-New Mexico. These results mirror actual developments. Powder River Basin coal has been moving into the midwest in substantial quantities since 1973. Shipments to Texas from Montana and Wyoming also began in 1973 and have been increasing since then.

The midwestern demand region is the swing region. The eastern pattern of consumption is relatively stable. Very small quantities of western coal reach eastern markets. There is some substitution between northern and southern Appalachian coal, but the eastern regions continue to burn coal produced east of the Mississippi River. Since this coal must meet sulfur restrictions, the eastern demand regions choose to scrub eastern coal. This is an important result, and one worth bearing in mind when analyzing recent policy initiatives aimed at reducing western coal production. The midwest is the only important area where the choice between eastern and western coals is sensitive to pollution regulation. The midwest thus presents the greatest sensitivity to coal policy in the United States.

Price Effect of Sulfur Constraints

As a result of sulfur constraints demand for low-sulfur coal increased and demand for high-sulfur coal decreased. We would expect therefore to see a decline in the price of high-sulfur coal and the creation of a premium for low-sulfur coal, which is in fact the result (see table 3.12). High-sulfur coal prices decline, and a large low-sulfur premium is created. The regional variation in this premium, however, is also large. In 1985, when the regulations reach 1.2 lb, the premium ranges from $.71 per million Btu (in 1985 dollars) in New England and the middle Atlantic to $.08 per million Btu in the mountain states. In 1979 dollars this is a range of 5.8 to 51 cents per million Btu or $.98 to $12.24 per ton of coal, respectively.[7]

The western consuming regions buy western low-sulfur coal regardless of sulfur regulations, since it is still the cheapest coal available. Low-sulfur western coal has a very elastic long-run supply curve, and increased production leads to only a slightly higher price than for high-sulfur coal. Adding transport cost to the east, however, results in a substantial premium in more eastern regions. The midwest pays a 17.4 cent (in 1979 dollars) per million Btu premium for western low-sulfur coal, or $3.83 per ton. Sulfur regulations raise the cost of coal in the east but have little effect in the west.

Sulfur pollution regulations are an important force affecting the industry. Acting alone they cause an estimated 34 percent increase in Powder River Basin coal production in 2000 and a 42 percent reduction in Illinois Basin coal production. Given the inelastic supply of low-sulfur, eastern coal, sulfur regulations are responsible for the large premia earned by low-sulfur coal in eastern markets. However, in reality sulfur pollution regulations are not acting alone. Factor price changes exert other influences on the coal industry.

3.3 Relative Factor Prices

Each mining activity—strip and deep—has been modeled as if factor proportions were fixed. That is, as relative costs of capital and wages change, we assume that the ratio of capital to labor within a mining unit is fixed.[8] However, deep mining is more labor intensive than strip mining. Furthermore the capital-to-labor ratios vary among regions for the same mining technique. Table 3.13 shows the estimated percentages of cost for labor, capital, and materials in 1975 in each region at the margin. These percentages are calculated from the cost functions of chapter 2. Factor price changes thus change the relative attractiveness of deep and strip mining at the margin.

The substitution of strip for deep mining (or vice versa) interacts strongly with depletion to produce important regional effects. As wages increase, for example, there is substitution away from deep mining to the

Table 3.12 Delivered cost of coal with constant factor prices, 1.2 lb SO_2 sulfur constraint (current \$/million Btu)

Year	Region								
	1	2	3	4J[a]	5	6	7	8J[a]	9
1980									
Low sulfur	1.62	1.50	1.37	1.05	1.55	1.35	1.28	0.74	1.24
High sulfur	1.37	1.27	1.26	1.01	1.33	1.18	1.25	0.74	1.22
1985									
Low sulfur	2.61	2.46	1.98	1.46	2.55	2.35	1.87	1.05	1.74
High sulfur	1.90	1.75	1.74	1.35	1.89	1.69	1.69	0.97	1.64
1990									
Low sulfur	3.51	3.32	2.59	1.92	3.45	3.18	2.58	1.51	2.37
High sulfur	2.60	2.41	2.31	1.73	2.66	2.27	2.18	1.24	2.11
1995									
Low sulfur	4.48	4.24	3.23	2.39	4.39	3.97	3.31	1.91	3.01
High sulfur	3.43	3.17	2.87	2.11	3.50	2.82	2.67	1.50	2.59
1999									
Low sulfur	6.15	5.88	4.32	3.23	5.88	5.29	4.49	2.66	4.05
High sulfur	4.56	4.26	3.80	2.71	4.66	3.74	3.51	1.94	3.41
1980–1999, average annual real increases									
Low sulfur	1.8%	2.0	0.7	0.6	1.8	2.0	1.3	1.5	0.9
High sulfur	1.0	1.1	0.5	neg.	1.3	0.8	neg.	neg.	neg.
1985–1999, average annual real increase[c]									
Low sulfur	0.8%	0.9	0.2	0.3	0.6	0.5	1.0	1.4[a]	0.7
High sulfur	0.9	1.1	0.2	neg.[b]	1.1	0.3	neg.[b]	neg.[b]	neg.[b]

[a]Region 4J and 8J are aggregates of several regions and the weight of each of these regions shifts over time. Thus, the aggregate price increase has no meaning.

[b]Prices in these regions and sulfur categories actually fall over time. This is because in earlier years there are constraints on shipments, as explained in chapter 2. By the end of the period these constraints are not binding and actual cost falls in real terms.

[c]Inflation, for simplicity, has been assumed to be 5.5 percent per year throughout the period.

Table 3.13 Fraction of total cost accounted for by labor, materials, and supplies (1977 factor prices)

Region	Capital	Labor	Materials
Strip production			
Northern Appalachia	0.52	0.21	0.26
Southern Appalachia	0.52	0.21	0.26
Midwest	0.51	0.18	0.31
Montana-Wyoming	0.39	0.16	0.45
Utah-Colorado	0.50	0.21	0.29
Arizona-New Mexico	0.48	0.15	0.36
Deep production			
Northern Appalachia	0.24	0.50	0.26
Southern Appalachia	0.24	0.50	0.26
Midwest	0.24	0.50	0.26
Montana-Wyoming	0.19	0.50	0.31
Utah-Colorado	0.19	0.50	0.31
Arizona-New Mexico	0.19	0.50	0.31

Note: Estimates from chapter 2.

less labor-intensive strip mining. This substitution is limited by depletion of strip reserves. The attempt to substitute strip mining means that strip reserves are depleted more quickly. This is serious in the midwest or east but, as we have seen, relatively unimportant in the west. Therefore we expect the net effect of rising wages is to accelerate mining in the west. When rising real wages are superimposed upon our previous cases, this is indeed what we observe.

To simulate the effects of recent labor cost developments, we impose actual wage developments upon the model. The recent history of labor costs has been one of increasing wages and decreasing productivity. Real capital costs have also been increasing, as can be seen by comparing the increase in the WPI for mining machinery to the national GNP deflator (see table 3.14).

To reflect the relative increase in labor costs, the model assumes the following scenario. The 1978 United Mine Workers contract calls for wage increases of 23.5 percent in 1978, 4.8 percent in 1979, and 4.6 percent in 1980. Wage increases are set equal to those percentages through 1980.[9] Given an assumed inflation rate of 5.5 percent, this implies a cumulative real increase of 18 percent or an annual average real increase of 5.6 percent. From 1981 to 1985 we assume a real increase of 2 percent

Table 3.14 Recent behavior of the mining machinery price index

Year	Mining machinery index	GNP deflator
1971	100.0	100.0
1972	103.0	104.1
1973	106.4	110.2
1974	126.4	120.8
1975	146.0	132.4
1976	186.3	139.3
1977		147.5
1978	219.4	158.4
1971–1978 average annual rate of increase	11.9%	6.8

Sources: Mining machinery index: U.S., Department of Labor, Bureau of Labor Statistics, *Wholesale Price Indices* (Washington, D.C.: Government Printing Office, various years). GNP deflator: *Economic Report of the President* (Washington, D.C.: Government Printing Office, January 1980), p. 206.

per year.[10] Thereafter real wages are held constant. Capital costs are assumed to increase in real terms at 2 percent per year until 1985. Thereafter real capital costs are held constant. The increase in real wages relative to capital costs by 1985 is 12.5 percent. Table 3.15 gives the results of this scenario for regional production.

The scenario assumes absolute increases in capital and wage costs and increases in wages relative to capital costs. The result is to reduce output in Montana and Wyoming by 7 percent, relative to the constant factor price case (table 3.15 compared to 3.11). At first glance this result seems paradoxical, as strip mining has become relatively cheaper. Yet total production in Montana-Wyoming decreases. In fact higher relative wages and higher capital costs have two effects. As we have explained, the rise in relative wages causes greater expansion of western coal at the margin. However, because of the higher cost of coal, demand declines and total output is significantly diminished. In the year 2000 total output is reduced by 8.4 percent. The percentage of output accounted for by Montana and Wyoming increases slightly from 55.7 to 56.8 percent, but total production in that region declines due to the demand effects. Thus the income effect swamps the substitution effect. We deal more extensively with demand effects later. At this point it suffices to point out the importance of an integrated analysis that captures demand effects. This scenario is interesting in a second way. It demonstrates that the effect of higher wages on production is small even in relative terms when added to the effect of

Table 3.15 Regional production under assumptions of changing relative prices and NSPS (million tons/year)

	Region	
Year	Northern Appalachia	Southern Appalachia
1980	201.2	245.2
1985	284.9	188.0
1990	350.0	166.1
1995	174.3	152.7
2000	186.1	175.9
Percent of national output in 2000	10.5	9.9

sulfur regulations. The incentive for western coal is created by sulfur pollution regulations; any additional increase in relative attractiveness of western coal due to wage and capital costs is small.[11]

3.4 Transportation Cost Increases

Not all factor price changes favor the development of western coal. Since western coal is located far from the main electrical load centers, the western coal industry is transportation intensive. Increases in transport cost work to the disadvantage of western coal. Since 1970 the real costs of coal transport have been increasing at a compound annual rate of 4 percent (see table 3.16). This trend, if continued, will have large effects on the regional allocation of output. We have been conservative and assume a trend of 3 percent annual real increase in transport costs until 1985, and constant thereafter, in order to analyze model sensitivity to current trends.

Table 3.17 shows the results of the moderate increase in transport costs. Montana and Wyoming produce 135 million tons less coal in 2000 with these higher transport costs. Again two forces produce this result. Total output declines because of the higher cost and reduced demand for coal. The decline is 77.3 million tons nationally in 2000, or 5 percent of output in the constant transport cost case. Montana and Wyoming futher suffer from the decline in the relative attractiveness of their coal.

The joint effect of these two forces in 1990 and 2000 can be seen in tables 3.18 and 3.19, which present distribution patterns with and without the transport cost increases. The comparison shows that in regions where Montana and Wyoming coal is still the only coal used, total demand

Midwest	Montana-Wyoming	Utah-Colorado	Arizona-New Mexico
136.8	193.0	13.5	61.3
79.1	345.3	31.9	79.7
142.3	690.1	48.4	55.8
299.6	1,025.8	50.6	37.7
295.2	1,006.7	47.0	60.2
16.6	56.8	2.7	3.4

Table 3.16 Recent behavior of rail cost indexes

Year	Implicit price deflator	AAR index	Year-to-year real increase in AAR index
1970	100	100	
1971	105	109	4.0%
1972	109	118	4.0
1973	116	133	6.3
1974	127	152	4.8
1975	139	173	4.4
1976	146	191	5.4
1977	155	208	3.7
1978	166	a	

Source: American Association of Railroads, Economics and Finance Department, *Indexes of Railroad Material Prices and Wage Rates, Railroads of Class 1,* Series QMPW-103 (Washington, D.C., April 25, 1979).

[a] According to Data Resources Incorporated, Transportation Service, the unit train rate increase from February 1978 to February 1979 was 14.8 percent.

Table 3.17 Regional coal production with transport cost increases of 3 percent per year to 1985 (million tons)

Year	Supply region					
	Northern Appalachia	Southern Appalachia	Midwest	Montana-Wyoming	Utah-Colorado	Arizona-New Mexico
1980	201.1	251.5	129.8	192.7	13.5	62.2
1985	264.3	196.9	84.4	328.7	31.7	78.4
1990	324.2	190.6	141.7	577.7	92.9	78.3
1995	327.5	207.2	105.8	848.9	71.2	58.1
2000	198.4	190.2	207.3	871.2	55.7	57.9

Note: This assumes sulfur standards reach 1.2 lb SO_2 by 1985 and wage and capital cost developments of table 3.16.

Basic Forces

Table 3.18 Comparison of coal distribution patterns in 1990 with and without transport cost increases (million tons)

Supply Region	Demand region					
	1	2	3	4J^a	5	6
Case: 1.2 lb SO$_2$ sulfur standard by 1985 and thereafter increasing real wage and capital costs; no increase in transport cost						
1	17.40	81.70	14.77	0.02	168.30	0.12
2	0.08	0.83	5.44	0.04	24.29	6.12
3	0	0	18.50	4.30	3.20	106.91
4	2.65	0	260.67	139.88	0	0.06
5	0	0	0.13	0.09	0	0
6	0	0	0	0	0	0
Total	19.77	82.54	299.51	144.33	195.80	113.20
Case: same as above except 3 percent real increase in transport rates 1978 to 1985						
1	19.15	81.37	14.77	0.02	140.76	0.12
2	0.08	0.83	5.44	0.04	48.84	6.12
3	0	0	18.50	4.30	3.20	106.31
4	0	0	244.05	135.51	0	0.06
5	0	0	0.13	0.09	0	0
6	0	0	0	0	0	0
Total	19.22	82.20	282.89	139.97	192.81	112.60

Note: Both cases assume declining sulfur standards through 1985 until 1.2 SO$_2$ is reached; UMW contract wages through 1980 and 3 percent real increase between 1980 and 1985, and constant thereafter; and 2 percent per year real increase in capital costs between 1978 and 1985. The table excludes metallurgical plus export coal.

[a] This region corresponds to the aggregate of demand regions 4 and 10.

[b] This region is the aggregate of demand regions 8, 11, and 12.

7	8J^b	9	10	11	12	Total
0	0	0	0	0	0	211.96
0	0	0	0	0	0	38.06
1.26	0	0	0.34	0	0	133.25
0	79.83	85.83	33.78	87.36	0	690.06
17.09	1.27	0	0	0	18.86	37.44
13.52	0.94	8.73	0	0	32.63	55.81
31.87	82.04	94.56	34.13	87.36	51.49	1,236.58
0	0	0	0	0	0	256.19
1.26	0	0	0	0	0	62.61
0	0	0	0.34	0	0	132.66
0	79.33	0	32.38	86.36	0	577.69
16.04	1.27	56.78	0	0	7.54	81.85
12.70	0.94	19.86	0	0	44.83	78.33
30.00	81.54	76.64	32.73	86.36	52.37	1,189.33

declines. This reflects the effect of higher prices on demand. This is most obvious in 2000. In several areas where Montana and Wyoming coal was used in the year 2000, there is substitution with coal from closer supply regions. The latter can be seen in demand regions 1, 2, 5, 6, 9, and 12: basically the east and west coasts. The former effect can be seen in regions 3, 4, 8, and 11: the midwest and southwest. In regions where there is substitution there are also clear declines in output.

3.5 Supply Developments

An industry unchanged with respect to relative factor prices and policy initiatives would have seen a substantial change in the underlying structure of production. Depletion alone would have caused the midwest and northern Appalachia to increase their relative share of production at the expense of southern Appalachia. The emergence of western demand regions would have caused a substantial rise in coal production west of the Mississippi River.

Table 3.19 Comparison of coal distribution patterns in 2000 with and without transport cost increases (million tons)

Supply Region	Demand region					
	1	2	3	$4J^a$	5	6
Case: no transport increase						
1	28.77	73.30	0	0	0	0
2	0	0	0	0	18.87	0
3	0	0	0	0	176.79	108.37
4	8.28	10.82	272.20	167.01	14.51	15.28
5	0	0	0	0	0	0
6	0	0	0	0	0	0
Total	37.04	84.12	272.20	167.01	210.18	123.65
Case: 3 percent per year real increase in transport rate 1978–1985						
1	33.63	80.71	0	0	0	0
2	0	0	0	0	30.81	2.35
3	0	0	0	0	183.93	113.40
4	0	0	247.36	130.26	0	5.91
5	1.86	0	0	0	12.26	0
6	0	0	0	0	0	0
Total	35.50	80.71	247.36	130.26	227.01	121.65

Note: Both cases assume declining sulfur standards through 1985 until 1.2 SO_2 is reached; UMW contract wages through 1980 and 3 percent real increase between 1980 and 1985, and constant thereafter; and 2 percent per year real increase in capital costs between 1978 and 1985. The table excluses metallurgical plus export coal.

[a] This region corresponds to the aggregate of demand regions 4 and 10.

[b] This region is the aggregate of demand regions 8, 11, and 12.

7	8J^b	9	10	11	12	Total
0	0	0	0	0	0	102.06
0	0	0	0	0	0	18.87
0	0	0	0	0	0	285.16
0	151.46	110.69	43.13	151.50	61.82	1,006.70
21.72	0	3.55	0	0	8.77	34.04
25.17	0	0.89	0	0	34.13	60.18
46.89	151.46	115.13	43.13	151.50	104.72	1,507.03
0	0	0	0	0	0	114.35
0	0	0	0	0	0	33.16
0	0	0	0	0	0	297.33
0	149.32	87.04	34.02	146.91	70.39	871.22
19.81	0	21.75	0	0	0	55.69
22.63	0	0	0	0	35.32	57.94
42.44	149.32	108.78	34.02	146.91	105.71	1,429.69

These two trends were influenced by a series of other developments. Our analysis indicates that sulfur pollution standards were responsible for creating a large premium for low-sulfur coal prices in the east. Where eastern low-sulfur coal must be burned, there is a large increase in the cost of coal. Sulfur regulations provided additional impetus for western coal development. The midwestern demand region, under pressure of sulfur pollution regulations, turns to western coal. The higher coal prices cause some cutback in total demand for coal, but the relative attractiveness of western coal increases sufficiently to counterbalance this trend.

The expected changes in relative factor prices have several additional counterbalancing effects. Higher real wages increase the relative attractiveness of western coal. Once pollution standards are introduced, however, the incremental effect on the west is small. This result has important implications. One might think that competition with western coal would lead the United Mine Worker's union to moderate wage demands. However, if our analysis is correct, there is little reason for them to do so. Lower wages alone would be unable to change shifting demand and depletion of low-sulfur eastern coal reserves.

The one development working in favor of eastern coal is the increase in transport rates. We have simulated an increase in rates due to escalating real costs. This pushes western coal out of the midwest market—its main swing market. To the extent that real costs escalate, the market position of western railroads will deteriorate; see Zimmerman and Alt [8].

These structural changes induced by factor price developments and sulfur regulations would cause substantial redistribution of factor payments, of mining employment, and of rents—all in favor of the west. Eastern coal producers, seeing this potential development, want to protect their markets. They are joined by the United Mine Workers union, which is dominated by eastern miners. The attempt to protect these markets and jobs have led to a series of policy initiatives aimed at income redistribution. Before turning to those issues, we need to understand the demand side of the market.

3.6 The Impact of Factor Price Changes on Demand

The combined models we are using allow us to investigate the impact of supply side developments on the demand for coal. Higher coal prices affect demand through several channels. Higher coal prices slow the rate of growth of energy demand. Electricity prices also rise, lowering the demand for electricity. At the same time, for any given level of demand, higher coal prices lead to substitution away from coal to other fuels.

Coal and Electricity Demand
The annual rate of growth in electricity consumption averaged above 7 percent from 1961 to 1973. Electric utilities became accustomed to forecasting future demand, using a 7 percent rate of growth year in and year out, but the increase in oil prices at the end of 1973 and in early 1974 shook this comfortable system. All fuel prices rose along with oil prices.[12] The price of electricity increased in tandem.[13] Growth in the demand for electricity was more than cut in half between 1973 and 1978. Since the recession of 1974 to 1975 further acted to reduce growth, current forecasts of long-term growth rates for electricity are in the neighborhood of 4 to 5 percent.[14] The estimated long-run demand for electricity in the model is consistent with these lowered expectations. Table 3.20 presents, for the four cases considered so far, the estimated rate of growth in electricity demand. As table 3.20 demonstrates, the model forecasts slow rates of growth in electricity demand, averaging about 4.5 percent for the entire period with small variations in each subperiod. More important, for our

Table 3.20 Average annual growth rate of electricity demand (percent)

Case	Period					
	1975 to 1980	1980 to 1985	1985 to 1990	1990 to 1995	1995 to 2000	Total
Constant factor prices no sulfur limit	4.7	5.0	4.7	4.2	4.1	4.52
Constant factor price 1.2 lb sulfur limit	4.6	4.9	4.5	4.3	4.3	4.48
Increase wages and capital prices 1.2 lb sulfur limit	4.6	4.7	4.4	4.2	4.4	4.47
Increased transport costs in addition to wages and capital increases	4.6	4.7	4.3	4.3	4.5	4.48

purposes, the final demand differs between the constant factor price case and the case including current trends in factor prices by slightly less than 1 percent. The forecasted rate of growth in demand is very low by historical standards and relatively unaffected by supply side developments.[15]

Coal production maintains a healthy growth rate even with the slow rate of growth in electricity demand. This is due to substitution away from other fuels in the electric utility sector. There are two aspects to the demand for coal in the electric utility sector: How does the increase in the price of coal affect the choice between different types of fuels? and How do utilities choose to meet the sulfur pollution regulations? Clearly there is a feedback between the cost of meeting regulations and the choice of fuel type.

Coal versus Nuclear Power
The most viable competitor for coal in the electric utility market over the period under consideration is nuclear power. Oil and gas are priced out of the electric utility market for new plant construction.[16] We are interested therefore in how basic forces operating in the coal industry affect the choice between nuclear power and coal. The main period of interest is toward the end of the twenty-five-year period, because the lags in construction are long. We have assumed a ten-year lag in the construction of nuclear plants and a five-year lag for coal. Recall that we constrain nuclear additions through 1988 to already announced plans. If a nuclear plant is not already in the pipeline, it cannot be available by 1988. We

have imposed one further set of constraints. Following the accident at the Three Mile Island station of the General Public Utilities Corporation, there is a great deal of uncertainty and caution with respect to the future growth of nuclear power. This has been modeled as a partial moratorium on nuclear construction. No new announcements for nuclear plants are permitted between 1978 and 1982. Because of the ten-year lag this means that no plants other than those in the pipeline today are built until after 1992. The choice between nuclear and coal therefore becomes important only in the last seven years of the period.

The choice is based on expected costs. Appendix table B.6 presents base-case expected capital costs for the various kinds of power-generating capacity. Given those costs, the model forecasts an increasingly important role for coal. Total capacity in place grows from 175.6 GW in 1975 to between 617 and 715 GW in 2000 (table 3.21). The electric utilty sector accounts, on a Btu basis, for 90 percent of total coal consumed in the year 2000. The increase in coal use comes at the expense of oil and gas. Nuclear capacity rises to between 334 GW and 407 GW in the year 2000. (The

Table 3.21 Generation capacity (GW)

	1975	1980	1985	1990	1995	2000
Case: constant factor prices, no sulfur standards						
FGD	0	0	0	0	0	0
Coal	175.587	299.642	320.434	473.263	627.689	715.66
LWR	34.744	64.46	122.812	125.381	199.534	352.39
Case: constant factor prices, 1.2 lb SO$_2$ standards						
FGD	0	0	0	175.656	186.613	222.677
Coal	175.587	299.642	318.545	291.556	425.314	497.618
LSW	34.744	64.46	122.812	125.381	209.076	377.795
Case: increased wages and capital costs, 1.2 lb SO$_2$ standard						
FDG	0	0	103.299	163.680	169.650	219.927
Coal	175.587	299.642	215.206	294.802	419.815	437.333
LWR	34.744	64.46	122.812	125.381	209.227	374.077
Case: increased wages, capital costs and transport rates, 1.2 lb SO$_2$ standard						
FGD	0	0	103.288	165.542	170.195	228.480
Coal	175.587	290.057	206.449	279.602	382.661	388.703
LWR	34.744	75.304	136.035	140.184	240.188	407.615

Note: FGD = coal with scrubbers; coal = coal without scrubbers; LWR = light-water reactors.

range in results is due to the effect of higher coal prices on the choice between nuclear power and coal. Increasing factor prices leads to higher coal prices and thus higher nuclear capacity.)[17]

Table 3.22 summarizes the delivered cost of coal by region. The increase in the cost of coal between our final case, which includes factor price increases, transport cost increases, and tight sulfur regulations, and the initial constant cost, no-sulfur regulation case is substantial. We can examine the impact of each set of changes using data in table 3.23.

We have already seen the effect of sulfur regulations. These caused an increase in price of from $.08 to $.71 per million Btu in 1985 (in 1985 dollars). The low end of the range was in the western states, the high end in New England. The effect of factor price changes is substantial, with the east realizing the highest increases. Wages are increasing relative to capital costs, thus driving up the relative cost of deep coal. The east relies more heavily on deep-mined coal. The increases in cost range from about 14 to 15 percent in the east to an average increase of 10 percent in the west in 1985.

The impact of the transport cost increase reverses this pattern. Western coal, on the average, is shipped over greater distances than eastern. The increase in transport costs leads to a delivered cost increase in the east of 3.8 to 7.3 percent. The western increases range from 4.5 to 10.1 percent. The highest increase is realized on midwestern low-sulfur coal costs. The midwest turns to western coal to meet sulfur regulations. This demand region thus pays the largest transport costs for low-sulfur coal. The eastern regions consume eastern coal, and the western regions consume western coal. The midwest, the swing region, pays the greatest relative transport costs.

The impact of these higher costs on the electric utility sector is to reduce installed coal capacity. By the year 2000 the effect is to reduce installed coal capacity by 14 percent in the final case compared to the constant-cost, no-sulfur regulations case (table 3.23).

The point is that in the later part of the period, there is considerable competition between nuclear power and coal. The nuclear capacity estimated here is below what government and private economists were forecasting five years ago for the year 2000. Even so, nuclear power assumes a large role in these scenarios. The total nuclear capacity estimated is far above what all forecasters now think will be achieved by the end of the century.[18] The pessimism about nuclear power reflects both regulatory hurdles and higher costs.[19] This has important implications: if nuclear power, because of public attitudes or regulatory difficulties, can-

Table 3.22 Delivered coal prices for selected regions, 1980 to 2000 (¢/million Btu in nominal $)

Demand	Region					
	2	3	5	7	8	9
1980						

Case: constant factor prices, no sulfur constraint

	2	3	5	7	8	9
Low sulfur	136.8	136.0	151.1	133.6	77.1	130.0
High sulfur	136.8	135.1	151.0	133.6	77.0	130.0

Case: constant factor prices, 1.2 lb SO_2 constraint

	2	3	5	7	8	9
Low sulfur	162.2	146.1	166.7	135.9	77.9	131.6
High sulfur	133.5	132.8	142.0	131.8	77.2	128.1

Case: increased wages and capital costs, 1.2 lb SO_2 constraint

	2	3	5	7	8	9
Low sulfur	172.7	150.1	177.2	140.1	81.3	136.6
High sulfur	141.2	140.4	149.1	136.4	80.5	132.7

Case: increased wages, capital costs and transport rates, 1.2 lb SO_2 constraint

	2	3	5	7	8	9
Low sulfur	175.8	157.3	180.4	145.8	82.9	141.2
High sulfur	143.9	143.1	152.4	141.4	82.2	137.3

1985

Case: constant factor prices, no sulfur constraint

	2	3	5	7	8	9
Low sulfur	183.5	179.8	203.0	176.4	100.1	170.8
High sulfur	183.5	179.8	202.9	176.4	98.3	171.2

Case: constant factor prices, 1.2 lb SO_2 constraint

	2	3	5	7	8	9
Low sulfur	246.0	173.7	255.1	186.9	105.13	174.0
High sulfur	174.6	197.8	188.7	169.1	97.3	164.2

Case: increased wages and capital costs, 1.2 lb SO_2 constraint

	2	3	5	7	8	9
Low sulfur	281.8	207.6	292.2	207.2	117.9	192.4
High sulfur	199.5	198.5	218.8	184.1	108.1	179.2

Case: increased wages, capital costs and transport rates, 1.2 lb SO_2 constraint

	2	3	5	7	8	9
Low sulfur	293.2	240.0	303.3	224.0	123.3	207.9
High sulfur	210.9	209.6	234.8	202.7	114.0	196.6

1990

Case: constant factor prices, no sulfur constraint

	2	3	5	7	8	9
Low sulfur	252.9	237.7	280	237.3	132.2	230.2
High sulfur	252.7	237.7	278.7	237.2	128.1	230.9

Case: constant factor prices, 1.2 lb SO_2 constraint

	2	3	5	7	8	9
Low sulfur	331.7	258.8	344.8	257.7	150.7	237.3
High sulfur	241.0	231.5	266.4	217.8	123.5	211.5

Table 3.22 (continued)

Demand	Region					
	2	3	5	7	8	9

Case: increased wages and capital costs, 1.2 lb SO_2 constraint

Demand	2	3	5	7	8	9
Low sulfur	379.8	273.0	384.6	280.1	168.2	256.5
High sulfur	277.4	258.0	302.7	237.4	137.5	231.0

Case: increased wages, capital costs and transport rates, 1.2 lb SO_2 constraint

Demand	2	3	5	7	8	9
Low sulfur	395.2	314.2	410.7	307.9	178.8	282.9
High sulfur	289.0	279.9	321.0	261.7	145.6	253.7

1995

Case: constant factor prices, no sulfur constraint

Demand	2	3	5	7	8	9
Low sulfur	353.6	318.0	240.8	331.1	173.3	307.4
High sulfur	354.5	317.9	231.9	332.9	181.8	307.0

Case: constant factor prices, 1.2 lb SO_2 constraint

Demand	2	3	5	7	8	9
Low sulfur	453.1	341.2	468.1	351.1	204.1	318.6
High sulfur	337.7	302.9	371.3	281.5	157.6	273.2

Case: increased wages and capital costs, 1.2 lb SO_2 constraint

Demand	2	3	5	7	8	9
Low sulfur	506.3	361.0	507.4	374.8	226.0	339.7
High sulfur	386.8	332.9	421.0	307.1	175.9	298.8

Case: increased wages, capital costs and transport rates, 1.2 lb SO_2 constraint

Demand	2	3	5	7	8	9
Low sulfur	541.1	414.0	558.0	425.8	242.4	385.4
High sulfur	403.9	365.2	445.8	338.9	186.4	328.4

2000

Case: constant factor prices, no sulfur constraint

Demand	2	3	5	7	8	9
Low sulfur	449.6	408.8	493.7	438.4	230.8	383.6
High sulfur	452.1	408.8	492.6	438.3	244.7	383.5

Case: constant factor prices, 1.2 lb SO_2 constraint

Demand	2	3	5	7	8	9
Low sulfur	588.0	432.2	587.8	448.8	265.6	404.7
High sulfur	426.1	380.5	465.9	350.8	193.5	341.1

Case: increased wages and capital costs, 1.2 lb SO_2 constraint

Demand	2	3	5	7	8	9
Low sulfur	640.6	459.7	640.1	475.9	295.1	433.8
High sulfur	489.1	409.4	528.8	388.3	211.7	377.7

Case: increased wages, capital costs and transport rates, 1.2 lb SO_2 constraint

Demand	2	3	5	7	8	9
Low sulfur	699.9	519.5	709.0	540.4	313.8	484.6
High sulfur	513.9	456.5	563.3	424.8	224.7	411.2

Table 3.23 Average national electricity prices under alternative scenarios (mills/kWh in current $)

1975	1980	1985	1990	1995	2000
27.4	36.4	48.4	64.2	84.7	111.5
27.3	36.9	49.6	66.9	87.5	113.9
27.3	37.2	51.1	68.4	89.3	113.6
27.3	37.1	51.9	69.2	89.9	115.6

Note: Annual inflation rate is assumed to be 5.5 percent for the entire period.

not expand to the levels seen here, the coal industry will assume a much larger role. Chapter 5 examines more pessimistic, but realistic, scenarios for nuclear power.

Choices among Coal-Using Technologies
How do utilities react to sulfur regulations? The capacity decisions shown in table 3.21 provide part of the answer; the coal price data in table 3.22 complete the picture. Stack gas scrubbing or fuel gas desulfurization becomes a viable technology. The decision to scrub is made in the absence of any regulation mandating use of Best Available Control Technology (BACT). Given the large increase in low-sulfur coal prices, it becomes economic in the east to scrub coal. West of the Mississippi, low-sulfur coal alone is favored to meet pollution regulations. The price data indicate the critical difference. In regions 1, 2, 5, and 6, the premium for low-sulfur coal is about $.70 per million Btu in 1985 ($.51 in 1979 dollars). The premium is only $.18 in region 7, $.06 in region 8, and $.10 in region 9. The midwest sees a premium of $.24, which causes it not to scrub. All this occurs once sulfur emission limits are introduced. Imposing the factor price changes on coal production hastens the introduction of scrubbing since the transport of low-sulfur coal is now more expensive. Thus scrubbing becomes economic over a large part of the U.S. and is used even in the absence of regulations mandating its use.[20]

Impact on Electricity Prices
Finally, the impact of the assumed factor price changes is small on the price of electricity. Coal prices increase on the order of 20 percent between the constant factor price scenario (with sulfur standards), and our final factor price increase case. The increase in the price of electricity is much smaller. In 1990, when nuclear constraints are binding, the electricity price increases by 4.6 percent. By the year 2000 greater substitutability

with nuclear power is possible, and the difference in electricity price declines to 1.4 percent.

Sulfur regulations alone cause an increase in the price of electricity. We can isolate the effect of sulfur regulations by comparing the scenario with no sulfur regulations and no factor price changes to the constant factor price case where sulfur regulations are imposed (see table 3.23). The price increase is greatest in 1990 and declines thereafter. In 1990 the increase is 4.2 percent, declining to 2.2 percent by the year 2000. In 1990 nuclear constraints are binding. The lower increase by the end of the period is due to the greater substitution with nuclear power at the end of the period. The combined effect of sulfur regulations and factor price increases is 7.8 percent in 1990 and 3.7 percent in 2000. In sum the real price of electricity increases in response to coal policy and basic forces operating in the coal industry.

The increases are moderated by two technological alternatives. The impact of sulfur regulations is moderated by the choice of scrubbing technology, and the impact of rising coal prices is mitigated by moving toward nuclear power at the margin.

Industrial Demand
The industrial sector accounts for 9 percent of present coal consumption. The model indicates that this will grow over time in response to changes in relative fuel prices.[21] Table 3.24 presents the model estimate for industrial coal demand in Btu. In parentheses we present the equivalent in tons of coal of 22 million Btu per ton. These figures exclude metallurgical and export coal, which are entered exogenously.

Table 3.24 tells an interesting story. The effect of pollution regulations is substantial. Increasing the cost of low-sulfur coal diminishes consumption of coal by the industrial sector a total of 18 percent by the year 2000 (compare line 2 with line 1). The combination of factor price increases and sulfur regulations reduces total consumption in 2000 by 54 percent. This suggests that the optimism expressed with regard to industrial coal consumption may not be warranted.[22] New technologies could change this picture. The introduction of fluidized-bed boilers could provide a way to burn the cheaper high-sulfur coal.[23] However, the penetration of new technologies is slow and not likely to substantially change the picture before 1990.

The second interesting question is: What fuel does the industrial sector consume? Coal demand grows slowly. As table 3.25 shows, industrial oil and gas consumption declines substantially in all scenarios, although the

Table 3.24 Industrial coal demand (10^{16} Btu)

1975	1980	1985	1990	1995	2000
Case: constant factor prices, no sulfur constraints					
0.2621	0.2280	0.2757	0.3529	0.4288	0.4996
(119)	(104)	(125)	(160)	(195)	(227)
Case: constant factor prices, 1.2 lb SO_2 constraints					
0.2621	0.2180	2.494	0.3059	0.3669	0.4253
(119)	(99)	(113)	(139)	(167)	(193)
Case: increased wages and capital costs, 1.2 lb SO_2 constraints					
0.2621	0.2154	0.2338	0.2758	0.3296	0.3790
(119)	(98)	(1061)	(125)	(1501)	(172)
Case: increased wages, capital costs and transport rates, 1.2 lb SO_2 constraint					
0.2621	0.21302	0.2193	0.2475	0.2873	0.3232
(119)	(97)	(100)	(113)	(131)	(147)

Note: Numbers in parentheses are million tons of 22 million Btu.

Table 3.25 Industrial oil and gas consumption (10^{16} Btu)

	1975	1980	1985	1990	1995	2000
Case: constant factor prices, no sulfur constraints						
Oil	0.2375	0.1898	0.1660	0.1566	0.1467	0.1358
Gas	0.7323	0.6970	0.5650	0.3974	0.3045	0.2598
Electricity	0.2852	0.4205	0.5782	0.7539	0.9291	0.1103
Case: constant factor prices, 1.2 lb SO_2 constraints						
Oil	0.2375	0.1913	0.1707	0.1678	0.1605	0.1483
Gas	0.7323	0.7015	0.5800	0.4253	0.3333	0.2840
Electricity	0.2852	0.4193	0.5738	0.7421	0.9210	0.1104
Case: increased wages and capital costs, 1.2 lb SO_2 constraints						
Oil	0.2375		0.1736	0.1739	0.1687	0.1556
Gas	0.7323		0.5898	0.4412	0.3504	0.2970
Electricity	0.2852	0.4191	0.5696	0.7372	0.9155	0.1110
Case: increased wages, capital costs and transport rates, 1.2 lb SO_2 constraint						
Oil	0.2374	0.1910	0.1739	0.1790	0.1739	0.1609
Gas	0.7321	0.7004	0.5912	0.4494	0.3615	0.3079
Electricity	0.2853	0.4211	0.5737	0.7400	0.9214	0.1123

decline is less severe the higher the coal price. Electricity expands the most in industrial fuel consumption. The average annual growth rate in electrical consumption in the industrial sector is 5.6 percent throughout the period. This compares to a 2.6 percent growth in direct consumption of coal. In reality this growth is a form of substitution of coal for other fuels in the industrial sector since coal plays such a large role in electrical generation.

3.7 User Costs

The model solves for the least-cost solution in each period. As such, it is a myopic view of resource scarcity. The reasons for this procedure were discussed in chapter 2. The solution, however, does not take into account future cost increases due to depletion. The question is whether a dynamic solution would be very different. In other words, if we assumed complete knowledge by producers and consumers about future cost increases, how different would our results be?

If costs were rising very rapidly, future run-ups in cost would be very important. Producers of low-cost reserves would earn scarcity rents. These rents arise since higher-cost deposits must be exploited in the future to satisfy demand. Prices would be greater than costs and neglect of these scarcity rents would be a grievous error. In essence this is the question of what has been called "user cost"; see Gordon [3]. Depletable resource markets would attribute a large rent to those resources for which costs will rise rapidly.

The scenarios analyzed here assume that these rents are small enough to be ignored. How bad is this assumption? We can do an *ex-post* analysis of this issue.

The rents arise because of depletion. The value of the rent can be shown to equal the present value of future cost increases.[24] Imagine our cumulative cost function as a step function, as was shown in figure 2.11. Each category of resource is available at a constant cost. Each step represents a different resource cost category. The user cost is equal to

$$\sum_{0}^{T} \Delta C(t) (1 + i)^{-t} \tag{3.1}$$

If we take a continuous approximation to this cost curve, the user cost is

$$\int_{0}^{T} \frac{dc}{dt} e^{-rt} dt. \tag{3.2}$$

Using equation (3.2), we could calculate a full dynamic equilibrium. We could solve the static model for the entire model period. Using equation (3.2), we would calculate a set of user costs associated with each supply curve. We could then solve the static model with those prices and recalculate user costs. The process would be repeated until convergence.[25] If user costs were very low, our initial solution would be very close to the ultimate outcome of this iterative process.

Here we attempt to calculate the first user-cost calculation. We take our solution to the linear program. We then use equation (3.2) and the results of the model for cost increases to examine the user costs associated with various supply functions. Tables 3.26 and 3.27 present the necessary cost information. The average rate of cost increase for all the scenarios for almost all the supply regions is small, less than 1 percent per year.

Let us approximate the cost path over time as

$$C(t) = C(0)e^{gt},$$

where $C(t)$ is cost in period t and g is average annual rate of growth associated with a given path.

Then

$$\frac{dC}{dt} = gC(0)e^{gt}$$

and

$$\text{User cost} = g \int_0^T C(0)\frac{dc}{dt} e^{gt}e^{-rt}dt$$

$$= \int_0^T gC(0)e^{(g-r)t}dt$$

$$= \frac{e^{(g-r)T} - 1}{g - r} \, gC(0).$$

If $g \neq r$, then user cost as a percentage of current cost is

$$\frac{g}{g-r}\left[e^{(g-r)T} - 1\right]. \qquad (3.3)$$

Using the results of the model for cost increases by type of coal, we estimate g. Substituting those values for g into equation (3.3) and taking interest rate equal to 10 percent yields a value for percentage error in our coal price calculations. These are presented in tables 3.26 and 3.27.

It is clear from our estimates that the error involved is small. Only low-sulfur eastern and midwestern coal is in inelastic supply. But, as we have pointed out, that coal disappears from steam coal markets early in the period. The production that does occur, occurs because of exogenous inputs to the model. We force a certain level of production to reflect metallurgical and export demand, which would occur regardless of price in this model.[26]

Inclusion of user cost in prices would raise Appalachian prices somewhat and have a very small effect in the midwest and Montana-Wyoming. Under sulfur regulations (table 3.27) the percentage rent in the lowest sulfur coal in Montana-Wyoming is $.75 per ton, which is small relative to the absolute level of rents in the east and to other factor price developments discussed in the following chapters. In sum, inclusion of user cost would exaggerate slightly our results. Appalachian output would be smaller in the early years, and output in the west and midwest would rise more rapidly. From the standpoint of policy analysis the exclusion of user cost from the solution leads to a bias on the conservative side, in keeping with the rest of our analysis.

In a sense this result appears to indicate that concern about coal depletion is misplaced. Coal prices would rise very slowly in response to depletion. However, this neglects the reason why coal prices rise slowly. They rise slowly because there is a large degree of interregional substitution. Price increases in one region lead to high rates of output in another, which slows the price increase. The disappearance of southern Appalachian low-sulfur coal occurs because western low-sulfur coal and scrubbed high-sulfur eastern coal replace it. Regional effects are in fact a key policy result. We would not know these without knowing which segments of the supply function are inelastic.

3.8 Basic Forces and Demand Policy Response

This chapter has taken a broad look at the effects of factor price changes on the demand for coal. We have seen that the electric utility industry will continue to dominate the demand for coal, and, as electric generation increases, coal use can be expected to increase substantially. Even though electric demand growth will be moderate by historical standards, the coal industry will expand at substantial rates as coal comes to account for a larger fraction of total fuel burned in the electric utility sector. Actual rates of growth in the industry are likely to be higher in the 1990s than forecast here if nuclear power continues to falter.

The second area in which coal will play an important role is the industrial sector. The high price of gas and oil will increasingly lead the

Table 3.26 Mine-mouth cost of coal ($/ton in nominal $) cumulative production, real cost increase, royalty for selected regions, without sulfur regulations

Supply region	Sulfur category		
	1	2	3
1			
1976 cost per ton	29.78	24.25	23.78
2000 cost per ton	129.95	101.10	102.68
Cumulative tons (millions)	153.39	223.65	533.27
Yearly real cost increase (%)	1.1	0.9	1.1
Royalty (%)	10.2	8.1	10.0
2			
1976 cost per ton	30.0	24.86	24.32
2000 cost per ton	160.68	105.09	100.56
Cumulative tons (millions)	1,691.6	1,340.00	790.86
Yearly real cost increase (%)	2.0	1.0	0.9
Royalty (%)	19.6	9.0	8.1
3			
1976 cost per ton	—	24.09	19.56
2000 cost per ton		136.27	79.75
Cumulative tons (millions)	0	75.75	73.84
Yearly real cost increase (%)	—	2.3	0.9
Royalty (%)	—	23.1	7.9
4			
1976 cost per ton	7.26	7.25	7.25
2000 cost per ton	24.27	24.13	24.22
Cumulative tons (millions)	3,071.08	2,464.0	1,930.0
Yearly real cost increase (%)	0.7	0.7	0.7
Royalty (%)	4.2	4.2	4.2

Note: Inflation is 5.5 percent per year, and the real discount rate is 10 percent. Royalties are calculated based on cost increases of state taxes, federal taxes, and union fees. The net sum of the latter two items is held fixed at $1.99 for the entire simulation period. The royalty is expressed as a percentage of price, including all taxes and fees. The lowest state severence tax in each region was used to calculate the cost net of the tax.

4	5	6	7	8
23.63	22.89	22.75	18.79	18.71
100.27	91.91	88.74	89.05	88.40
626.56	589.34	573.49	930.28	3,532.45
1.0	0.8	0.6	1.6	1.6
9.0	7.0	5.2	14.8	14.7
24.05	23.30	23.21	19.76	19.86
100.55	100.73	99.87	100.28	100.74
347.57	276.54	126.05	109.46	151.34
0.9	1.1	1.1	1.9	1.9
8.1	10.0	10.0	18.1	18.1
19.44	19.53	19.56	19.61	19.48
79.71	79.49	79.44	79.48	79.33
190.97	196.82	262.5	360.93	6,975.6
0.9	0.9	0.9	0.9	0.9
7.9	7.9	7.9.	7.9	7.9
7.25	7.25	7.25	7.25	7.25
24.21	24.20	24.16	24.16	24.16
847.0	189.0	32.83	25.52	34.26
0.7	0.7	0.7	0.7	0.7
4.2	4.2	4.2	4.2	4.2

Table 3.27 Mine-mouth cost of coal for selected regions, with sulfur regulations of 1.2 lb SO_2 per million ($/ton in nominal $)

Supply	Sulfur category		
region	1	2	3
1			
1976 cost per ton	29.78	24.25	23.78
2000 cost per ton	144.81	111.6	102.48
Cumulative tons (millions)	217.0	442.0	539.0
Yearly real cost increase (%)	1.6	1.4	1.1
Royalty (%)	15.0	13.0	10.0
2			
1976 cost per ton	30	24.86	24.32
2000 cost per ton	164.37	109.23	102.33
Cumulative tons (millions)	1,767.9	1,917.0	865.0
Yearly real cost increase (%)	2.1	1.2	1.0
Royalty (%)	21.1	11.0	9.0
3			
1976 cost per ton	—	24.09	19.56
2000 cost per ton	—	135.59	87.42
Cumulative tons (millions)	0	75.75	115.98
Yearly real cost increase (%)	—	2.3	1.3
Royalty (%)	—	23.1	11.8
4			
1976 cost per ton	7.26	7.25	7.25
2000 cost per ton	29.5	23.84	20.73
Cumulative tons (millions)	11,767.71	1,190.0	73.56
Yearly real cost increase (%)	1.6	5.7	0
Royalty (%)	10.3	3.5	0

Note: Inflation is 5.5 percent per year, and the real discount rate is 10 percent. Royalties are calculated based on cost increases of state taxes, federal taxes, and union fees. The net sum of the latter two items is held fixed at $1.99 for the entire simulation period. The royalty is expressed as a percentage of price, including all taxes and fees. The lowest state severance tax in each region was used to calculate the cost net of the tax.

4	5	6	7	8
23.63	22.89	22.75	18.79	18.71
99.77	91.45	82.32	82.35	82.42
626.0	589.0	270.0	729.0	2,812.0
1.0	0.7	0.3	1.2	1.3
9.0	6.1	2.6	10.7	11.7
24.05	23.30	23.21	19.76	19.54
93.8	93.75	93.44	93.36	93.26
248.0	208.0	93.65	93.2	131.14
0.7	0.7	0.3	1.2	1.3
6.1	6.1	2.6	10.7	11.7
19.44	19.53	19.56	19.61	19.48
75.06	73.08	72.59	72.64	73.02
122.08	95.04	116.02	163.88	3,250.0
0.5	0.5	0.5	0.5	0.5
4.2	4.2	4.2	4.2	4.2
7.25	7.25	7.25	7.25	7.25
20.71	20.67	20.77	20.87	20.73
24.76	3.32	1.63	2.09	1.29
0	0	0	0	0
0	0	0	0	0

industrial sector to turn to direct burning of coal and, more important, indirect burning of coal through the increased consumption of electricity.

Finally, substitution between technologies plays an important role. The expansion of coal use is impressive. However, the basic economics of the utility industry call for a substantial amount of nuclear capacity. The effect of increased coal prices on electricity prices is moderated by the expansion of nuclear power. Similarly the effect of sulfur pollution regulations is moderated by the adoption of stack gas scrubbing devices.

The basic economic forces analyzed are forcing a dramatic restructuring of coal production and consumption. New supplying areas assume a great importance under the pressure of sulfur regulations and factor price changes. The electric utility sector also faces important changes. Coal and nuclear power are called upon in many areas of the United States to replace previous reliance on gas and oil. New relationships between producing and consuming regions are created by these basic forces.

We have examined the logical implications of depletion, antipollution regulations, and factor price changes. These forces create changes that are not acceptable to large segments of the U.S. population. Environmentalists do not want to see an expansion in western strip mining; antinuclear groups do not want to see the expansion of nuclear power; and eastern coal producers and miners do not want to see the decline of eastern coal markets. Some groups are calling for the abolition of environmental restrictions to increase coal use and limit oil imports. But all are independently attempting to change policy to affect the course of the industry. Each of the policy actions proposed by these groups entails costs. What are these costs and what do we gain, in terms of the objectives of these groups, by enacting these policies? What are the trade-offs involved in policy choice? It is to those issues that we now turn.

References

1. Baughman, M., G. Rozanski, and M. B. Zimmerman. "Coal Model Integration." MIT Energy Laboratory working paper (forthcoming).

2. Edison Electric Institute. *Statistical Yearbook,* 1973 and 1974.

3. Gordon, R. L. "A Reinterpretation of the Pure Theory of Exhaustion." *Journal of Political Economy,* 75 (1967): 274–286.

4. Herfindahl, O., and A. Kneese. *Economic Theory of Natural Resources.* New York: Merrill, 1974.

5. Modiano, E. *Normative Models of Depletable Resources.* Ph.D. dissertation. Sloan School of Management, MIT, 1978.

6. U.S., Department of Energy, Energy Information Administration. *Annual Report to Congress 1978,* vol 3. Washington, D.C.: Government Printing Office, 1979.

7. White, D. E. *The Application of Mathematical Programming Techniques to the Integration of a Large Scale Energy Model.* Ph.D. dissertation. Department of Civil Engineering, Water Resources Program, MIT, 1980.

8. Zimmerman, M. B., and C. Alt. "The Western Coal Tax Cartel." MIT Energy Laboratory working paper (forthcoming).

9. Zimmerman, M. B., and R. P. Ellis. "What Happened to Nuclear Power?" MIT Energy Laboratory working paper 80-002WP, January 1980.

4   The Economics of Environmental Trade-Offs

In the United States energy policy increasingly connotates policy toward oil imports. The success of individual actions is judged by how much oil is saved. Congress is criticized because it passes laws that make it difficult to substitute coal for oil. The EPA is coming under increasing fire because it does not serve energy policy. In sum energy policy seems to be viewed as actions aimed at reducing oil consumption and imports.

This approach to energy policy puts the cart before the horse. We should have no energy objectives. The end we seek to advance is our national welfare. Rightly or wrongly, in the minds of our policy makers reduced imports strengthen our political independence. But national security is only one of many elements of national welfare. Other aspects of national welfare we are concerned with are economic growth, correcting environmental damage, and income distribution. We take actions within the energy sector aimed at each of these goals. But often the pursuit of one goal limits our ability to satisfy others. For example, environmental regulations increase oil imports by making coal burning too costly. Price controls on oil and gas might have favorable implications for income distribution that please a majority of the electorate, but they also increase imports. These measures may be seen as inconsistent but they are inconsistent only if one rejects environmental goals or income distribution goals as legitimate objectives of public policy. We pursue many objectives and trade off among them. For example, we are willing to increase imports if the environmental damage we thereby avoid is great. We are willing to end controls when their inefficiencies become large. What is important then is what we gain in terms of one objective when we relax in our pursuit of another.

The argument is that we do not have a comprehensive and independent energy policy. Instead we have a national security policy, an environmental policy, and so forth.

4.1 Sulfur Emission Standards and the Price of Coal

Chapter 3 examined the effects of a 1.2 lb SO_2 sulfur emission standard. There we isolated and analyzed the impact of sulfur regulations in the absence of factor price changes. This chapter expands that analysis first by examining sulfur regulations in conjunction with factor price developments and then a range of sulfur standards in order to understand the trade-offs involved in sulfur pollution regulation.

There are several dimensions to the assessment of sulfur emission standards. How do lower emission standards affect the price of both low- and high-sulfur coal? How will they affect the price of electricity, and the regional distribution of output? The goal here is to assess the terms of the trade-off. We must pay a price to lower air pollution. Who pays and how much do they pay?

The price paid for low-sulfur coal depends crucially upon the requirements of antipollution regulations. Without antipollution regulations low-sulfur coal would earn a small premium above high-sulfur coal. Users would be willing to pay for the less corrosive nature of low-sulfur coal. Sulfur reacts with moisture during combustion to form sulfuric acid, which corrodes boilers and leads to higher maintenance costs. To save those costs, users would pay a premium for lower-sulfur content. This premium, however, would be small and equal to the savings in maintenance cost.

The introduction of sulfur emission regulations greatly increases the demand for low-sulfur coal. The inelastic supply in the east, when confronted with an increased demand, leads to a high premium for low-sulfur coal. As eastern low-sulfur prices rise, eastern regions can import western low-sulfur coal. Since the elasticity of supply is large in the west, this will not lead to a great premium for coal in Montana at the mine mouth. The cost of transporting western coal eastward is the premium that eastern regions must pay for the low-sulfur product. This suggests that the effects of pollution regulations will be quite different from region to region.

Pollution regulations interact with other developments in the industry. If transport costs increase, for example, the premium that eastern regions must pay for western coal also increases. We therefore conduct this analysis on the basis of a common set of assumptions about factor price developments. These are the assumptions of the final case in the preceding chapter. Capital costs, wages, and transport costs are all assumed to increase in real terms until 1985 and are constant thereafter.[1]

The impact various levels of emissions have on the price of low-sulfur

The Economics of Environmental Trade-offs

coal is shown in table 4.1. The lowest emission level is set at 8 lb of SO_2 per million Btu. This is equivalent to a limit of 4.4 percent sulfur (by weight) in 1 ton of midwestern coal—in effect no sulfur limit at all. The intermediate level is 4 lb SO_2 per million Btu, or 2.2 percent sulfur by weight. The tightest level is the current standard of 1.2 lb SO_2 or 0.66 percent by weight.

It is clear from the table that the effect of reducing sulfur emissions is highly nonlinear. Cutting emissions in half from 4.4 to 2.2 lb has only a small effect on low-sulfur coal prices. The effect in 1980 ranges from a 5 percent increase in low-sulfur coal prices in New England to no effect at all in the western demand regions. In 1990 the low-sulfur coal price increase is 6 percent in New England, and there is still no effect in the western states. This pattern is exactly what we would expect. The east, in

Table 4.1 Delivered price of low-sulfur coal (current \$/million Btu)

Case	Demand region		
	1	2	3
1980			
8 lb SO_2 per million Btu	1.47	1.35	1.35
4 lb SO_2 per million Btu	1.55	1.44	1.37
1.2 lb SO_2 per million Btu	1.71	1.59	1.44
1985			
8 lb SO_2 per million Btu	2.21	2.04	2.00
4 lb SO_2 per million Btu	2.33	2.15	2.08
1.2 lb SO_2 per million Btu	2.82	2.64	2.21
1990			
8 lb SO_2 per million Btu	3.08	2.84	2.70
4 lb SO_2 per million Btu	3.25	3.02	2.77
1.2 lb SO_2 per million Btu	3.96	3.73	2.98
1995			
8 lb SO_2 per million Btu	4.29	3.99	3.59
4 lb SO_2 per million Btu	4.43	4.12	3.65
1.2 lb SO_2 per million Btu	5.35	5.05	3.92
2000			
8 lb SO_2 per million Btu	5.79	5.39	4.86
4 lb SO_2 per million Btu	5.99	5.61	4.80
1.2 lb SO_2 per million Btu	7.42	7.01	5.20

[a]The J denotes the census region aggregate of demand areas.

order to meet sulfur regulations, turns to Appalachian low-sulfur coal. Given the relatively inelastic supply, the price rises. The west consumes western coal regardless of regulations. It is the west's cheapest source of coal. The midwest is the most important of the regions that consume more western coal under sulfur regulations. However, the supply of western coal is so elastic that incremental production due to sulfur regulation does not increase its price. The results for the year 2000 confirm this.

As we saw in chapter 3, lowering emissions to 1.2 lb SO_2 per million Btu burned has a much more significant impact. In this range the cost of lowering sulfur emissions is much higher. The initial reduction to 4 lb raised the low-sulfur coal price in 1990 in New England by 6 percent, but the reduction in emissions to 1.2 lb SO_2 raises the 1990 low-sulfur coal cost in New England by an additional 21 percent.[2] The effects are equally

$4J^a$	5	6	7	$8J^a$	9
1.10	1.50	1.33	1.33	0.77	1.29
1.10	1.55	1.34	1.33	0.77	1.29
1.10	1.63	1.43	1.34	0.77	1.30
1.52	2.26	1.97	1.95	1.09	1.89
1.58	2.32	2.07	1.95	1.08	1.89
1.61	2.75	2.50	2.03	1.13	1.92
2.01	3.14	2.64	2.68	1.47	2.60
2.10	3.22	2.89	2.67	1.44	2.60
2.18	3.87	3.55	2.90	1.64	2.66
2.63	4.38	3.53	3.67	1.99	3.53
2.77	4.38	3.83	3.69	1.94	3.55
2.87	5.22	4.81	3.99	2.28	3.65
3.49	5.92	4.77	5.32	2.84	4.66
3.65	5.95	5.14	5.35	2.77	4.69
3.83	7.10	6.37	5.42	3.15	4.85

dramatic elsewhere in the east. Middle and south Atlantic low-sulfur coal costs both rise by 23 percent. The impact is generally much more moderate in the west, ranging from 2.5 percent in the Pacific region to 8.4 percent in Texas. The impact in region 8 is misleading. Recall that census region 8 is an aggregate of three separate regions in the coal model, extending from the border with Canada to the border with Mexico. Under tighter sulfur regulations the southern part of that region must shift from local Arizona-New Mexico coal to more expensive Utah-Colorado coal. The resulting cost increase is large in percentage terms because of the initially low cost of coal in region 8.

The impact of tighter sulfur emissions on high-sulfur coal prices is more moderate for all regions. Table 4.2 presents the high-sulfur coal price under the three emissions levels for 1990.

Table 4.2 Delivered price of high-sulfur coal (current $/million Btu)

	Demand region		
Case	1	2	3
1980			
8 lb SO_2 per million Btu	1.47	1.35	1.35
4 lb SO_2 per million Btu	1.45	1.33	1.33
1.2 lb SO_2 per million Btu	1.44	1.33	1.32
1985			
8 lb SO_2 per million Btu	2.21	2.04	2.00
4 lb SO_2 per million Btu	2.17	1.99	1.98
1.2 lb SO_2 per million Btu	2.12	1.94	1.93
1990			
8 lb SO_2 per million Btu	3.08	2.84	2.70
4 lb SO_2 per million Btu	2.96	2.73	2.65
1.2 lb SO_2 per million Btu	2.95	2.72	2.64
1995			
8 lb SO_2 per million Btu	4.29	3.99	3.60
4 lb SO_2 per million Btu	4.04	3.74	3.49
1.2 lb SO_2 per million Btu	4.11	3.80	3.46
2000			
8 lb SO_2 per million Btu	5.78	5.39	4.86
4 lb SO_2 per million Btu	5.51	5.11	4.60
1.2 lb SO_2 per million Btu	5.54	5.14	4.58

[a]The J denotes the census region aggregate of demand areas.

The following points emerge from this analysis. The impact of emissions standards on the price of coal is highly nonlinear. The bulk of the increase in coal prices occurs when standards are tightened from 4 to 1.2 lb SO_2 per million Btu burned. The price impact is highest for eastern low-sulfur coal and minimal for both high-sulfur and western low-sulfur.

In essence we need keep in mind three separate supply functions for coal: eastern high-sulfur, eastern low-sulfur, and western low-sulfur. One of these curves, eastern low-sulfur, is inelastic. These three supply curves lie behind the basic policy choices facing the country. Sulfur regulations cause a movement to western coal in order to avoid high cost increases in eastern coal. This entails more strip mining and thus a different environmental cost. It also entails higher delivered coal prices. These costs have implications for both the coal industry and the U.S. economy.

$4J^a$	5	6	7	$8J^a$	9
1.06	1.50	1.33	1.33	0.77	1.29
1.06	1.48	1.31	1.33	0.77	1.29
1.06	1.40	1.24	1.31	0.77	1.28
1.52	2.26	1.97	1.95	1.09	1.89
1.52	2.22	1.96	1.94	1.09	1.89
1.51	2.10	1.86	1.88	1.07	1.83
2.01	3.14	2.64	2.68	1.47	2.60
2.02	3.03	2.60	2.68	1.47	2.60
1.98	3.02	2.59	2.49	1.39	2.41
2.74	4.38	3.53	3.69	1.92	3.53
2.66	4.11	3.42	3.69	2.00	3.54
2.54	4.20	3.40	3.22	1.77	3.12
3.62	5.92	4.77	5.32	2.74	4.66
3.51	5.61	4.52	5.34	2.88	4.69
3.29	5.65	4.50	4.24	2.25	4.12

4.2 The Effects of Alternative Sulfur Emission Levels on Fuel Choice

We have assumed that industrial plants must burn low-sulfur coal to meet pollution requirements. The assumption is justified because of economies of scale in scrubber technology and the relatively small size of industrial boilers.[3] The rise in the price of low-sulfur coal has a moderate impact on industrial use of coal. Table 4.3 shows industrial use of coal for the three emission standards, excluding production of metallurgical and export coal.[4] The effect of sulfur pollution limits of 1.2 lb is a 14.5 precent reduction in industrial use of coal by 2000. The growth in industrial use is cut from a 2.8 percent annual rate of growth to 2.2 percent per year.

Given the oil prices considered here, the economics of fuel choice yield only a moderate growth in industrial coal consumption. This is true even without tight sulfur emission limitations. Sulfur emission limitations further erode growth in this sector. The U.S. government has at various times made forecasts of industrial coal consumption. Early energy plans

Table 4.3 Industrial use of coal (tons of 22 million Btu)

Case	Demand region		
	1	2	3
1980			
8 lb SO$_2$ per million Btu	1.8	16.7	30.4
4 lb SO$_2$ per million Btu	1.7	15.8	30.1
1.2 lb SO$_2$ per million Btu	1.6	14.8	29.4
1985			
8 lb SO$_2$ per million Btu	3.3	15.6	25.3
4 lb SO$_2$ per million Btu	3.1	14.4	24.5
1.2 lb SO$_2$ per million Btu	2.5	11.8	23.3
1990			
8 lb SO$_2$ per million Btu	5.0	16.7	24.8
4 lb SO$_2$ per million Btu	4.5	15.1	24.0
1.2 lb SO$_2$ per million Btu	3.4	11.3	22.2
2000			
8 lb SO$_2$ per million Btu	6.9	18.9	28.7
4 lb SO$_2$ per million Btu	6.4	17.4	28.4
1.2 lb SO$_2$ per million Btu	4.7	12.3	26.4

Note: Excludes metallurgical and exports.
[a]The J denotes the census region aggregate of demand areas.

of the Carter administration saw substantial expansion of industrial coal use. But the results here suggest that unless fuel choice is mandated by government regulations, coal is not going to play a very major role in the industrial sector over the time period considered here.[5]

Sulfur regulations have an important effect on the electric utility industry's choice of emission control technology. Under the pressure of sulfur regulations, flue gas desulfurization becomes a viable technology (see table 4.4). However, the imposition of a 4 lb standard is not sufficient incentive to the introduction of scrubbing. Utilities meet pollution requirements with coal low enough in sulfur to be burned without scrubbers. As we saw in chapter 3, scrubbing becomes an attractive alternative only under pressure of 1.2 lb sulfur regulations. By the year 2000 scrubbing will be applied to 37 percent of total utility coal capacity. The regional pattern of scrubbing is as would be expected (see table 4.5). The Atlantic seaboard (regions 2, 5, 6), with the exception of New England and the east south central region, finds scrubbing to be the economic

$4J^a$	5	6	7	$8J^a$	9	Total
9.1	12.7	9.5	7.5	8.1	5.7	101.5
9.1	12.4	9.5	7.5	8.0	5.8	99.9
9.1	11.8	9.2	7.5	8.0	5.8	97.2
12.5	13.3	8.7	10.9	13.5	8.1	111.2
12.5	12.8	8.5	10.9	13.5	8.2	108.4
12.5	11.1	7.4	10.6	13.2	8.1	100.5
16.6	15.6	9.3	12.8	20.2	10.7	131.7
16.7	15.0	8.6	12.9	20.3	10.8	127.9
16.5	12.2	6.8	12.2	18.9	10.8	114.3
23.9	20.3	11.4	14.1	36.1	15.4	175.7
23.9	20.2	10.0	14.2	36.0	15.5	172.0
23.6	15.3	7.1	13.8	30.0	15.7	148.9

choice. New England chooses no coal by 1985, and therefore scrubbing does not appear there.

The scrubbing capacity chosen through 1985 is retrofit capacity. Scrubbers are installed on existing plants that have many years of useful life. Only in the post-1985 period are new plants with scrubbers built. The new plants with scrubbers are built predominantly in the east. Areas west of regions 1, 2, 5, and 6 find the lowest-cost compliance strategy to be use of the low-sulfur coal. The exceptions are in the mountain and west south central regions where some scrubbing is introduced in the year 2000. It seems strange that these regions, close to western coal, choose scrubbing, but in fact this is consistent with what we saw earlier with regard to sulfur premia. Region 7, the south central, and parts of region 8 choose Arizona-New Mexico coal under a no-sulfur regulation regime. When tight standards are imposed, these regions must switch suppliers, and a sulfur premium is created. The premium causes the adoption of scrubbing by the year 2000. This is an important result that has implications for policy with respect to BACT. We will come back to this point in subsequent analysis.

Table 4.4 Electrical generation capacity by type of plant (GW)

Case	FGD[a]	Gas	Oil	Coal	Nuclear	Other	Total
1980							
8	0	145.5	35.4	299.6	64.5		683.1
4	0	145.5	35.4	299.6	64.5		683.1
1.2	0	145.5	35.4	299.6	64.5		683.1
1985							
8	0	130.0	59.6	320.0	122.8		780.2
4	0	130.0	59.0	319.4	122.8		779.0
1.2	105.7	130.0	58.0	212.6	122.8		777.3
1990							
8	0	113.0	85.2	455.4	125.4		943.3
4	0	113.0	85.2	452.9	125.4		941.4
1.2	168.2	113.0	83.7	281.6	125.4		935.6
2000							
8	0	80.0	74.9	625.11	414.4		1,383.7
4	0	80.0	74.9	625.4	410.7		1,380.1
1.2	233.0	80.0	73.5	391.7	392.1		1,363.9

[a]FGD = coal with flue gas desulfurization.

Sulfur regulations have little effect on total coal capacity installed. The surprising result is that sulfur regulations lead to a slight reduction in nuclear capacity. This arises for the following reason. Nuclear capacity is constrained through 1992. In years after 1993 it is the economic choice for large areas of the United States. Sulfur regulations raise the cost of electricity and therefore lower the demand for electricity. The 1.2 lb standard yields a total demand for electricity in 2000 that is 1.4 percent below the 8 lb case (see table 4.6). Fewer plants are built. The cutback in total capacity comes almost entirely from the nuclear power sector. This is a phenomenon akin to the effect on the nuclear industry of rising oil prices.

Table 4.5 Scrubbing capacity by region (GW)

Case	Demand region								
	1	2	3	4J	5	6	7	8J	9
1980									
8 lb SO₂	0	0	0	0	0	0	0	0	0
4 lb SO₂	0	0	0	0	0	0	0	0	0
1.2 lb SO₂	0	0	0	0	0	0	0	0	0
1985									
8 lb SO₂	0	0	0	0	0	0	0	0	0
4 lb SO₂	0	0	0	0	0	0	0	0	0
1.2 lb SO₂	0	0	0	0	0	0	0	0	0
1990									
8 lb SO₂	0	0	0	0	0	0	0	0	0
4 lb SO₂	0	0	0	0	0	0	0	0	0
1.2 lb SO₂	0	17.6	0	0	50.9	37.2	0	0	0
2000									
8 lb SO₂	0	0	0	0	0	0	0	0	0
4 lb SO₂	0	0	0	0	0	0	0	0	0
1.2 lb SO₂	14.2	32.6	0	0	83.8	48.5	27.0	26.8	0

Table 4.6 Total demand for electricity (million MWh)

Standard	1995	2000	Percent decline in 2000 relative to 8 lb standard
8 lbs SO₂	4,861.1	6,006.4	—
4 lb SO₂	4,850.4	5,993.2	0.2
1.2 lb SO₂	4,746.8	5,922.9	1.4

When oil prices jumped precipitously in 1974, the comparative economics of nuclear power improved substantially. Nevertheless, the demand cutbacks that followed the higher energy prices caused a disproportionate reduction in forecasted nuclear capacity.

Effects on Regional Production

Sulfur regulations appear to have a relatively small effect on the total quantity of coal consumed nationwide. We have seen, however, that the regional allocation of coal output is quite sensitive to sulfur standards. This is true in the cases considered here. Lowering standards from 8 to 4 lb SO_2 per million Btu increases production in Montana and Wyoming by about 100 million tons in the year 2000. The lowering to 1.2 lb increases Montana and Wyoming output by an additional 200 million tons, to 870 million tons in the year 2000. The imposition of the tight standard then is responsible for over a 50 percent increase in the simulated output of Montana-Wyoming coal (see table 4.7).

Table 4.7 Regional coal production under alternative sulfur emission levels (million tons)

Case	Supply region					
	1	2	3	4	5	6
1980						
8 lb SO_2	323.0	193.0	117.5	131.2	13.5	62.2
4 lb SO_2	263.8	216.6	135.2	157.7	13.5	62.3
1.2 lb SO_2	201.1	251.5	129.8	192.7	13.5	62.2
1985						
8 lb SO_2	234.0	192.7	251.8	204.6	22.1	88.6
4 lb SO_2	255.9	228.4	136.6	263.3	22.1	89.0
1.2 lb SO_2	264.3	196.9	84.4	328.7	31.7	78.4
1990						
8 lb SO_2	344.0	195.4	349.6	337.8	68.9	102.6
4 lb SO_2	298.9	273.9	218.9	447.7	71.6	101.0
1.2 lb SO_2	324.2	190.6	141.7	577.7	92.9	78.3
2000						
8 lb SO_2	240.8	166.0	490.6	574.1	53.0	99.0
4 lb SO_2	311.6	287.8	220.0	678.4	33.3	99.4
1.2 lb SO_2	198.4	190.2	207.3	871.2	55.7	57.9

Note: Includes metallurgical and export.

The responses of demand regions to low-sulfur regulations are varied. In the east, regions 5 and 6 switch from midwestern coal to southern Appalachia coal (see table 4.8). The other eastern regions—New England and the mid-Atlantic—stay committed to northern Appalachia coal. Likewise for western regions, which continue to consume western coal with or without pollution regulations. The greatest changes take place in the midwestern consuming area. With no-sulfur pollution regulations it consumes midwestern coal exclusively. With a 1.2 lb SO_2 standard it diverts a substantial fraction of its coal consumption (46 percent) to Montana-Wyoming. The midwest is a swing area in U.S. coal production. Sulfur regulations and policies analyzed next indicate the extreme sensitivity of consumption patterns in this region.

A rise of this magnitude in western low-sulfur coal production is awesome. The 1.2 lb SO_2 standard calls for a 7.8 percent average annual increase in Montana and Wyoming production. Massive amounts of infrastructure would be required. Since all this coal would be strip mined, this scenario calls for an enormous expansion in the amount of strip mining in the United States. Given the political opposition this entails, it is not surprising to expect policies aimed at reversing this trend. In fact the main thrust of U.S. coal policy is aimed at reversing this movement to western coal.

Since sulfur regulations substantially increase the price of low-sulfur coal, of course raising the price of electricity, the net impact on the price of electricity will be reduced if utilities can substitute scrubbers and high- for lower-sulfur coal. Given the analyses of the previous sections, the reduction from 8 to 4 lb SO_2 has little impact on the price of electricity. The national average price increase due to a 4 lb standard is 1.2 percent in 1985, 1.4 percent in 1990, and 0.7 percent in 2000 (see table 4.9). The decline at the end of the period is again due to the relaxation of constraints on nuclear power, which serves to mitigate the impact of higher coal prices.[6] Regional impacts are highest in the east and midwest and lowest in the west (see table 4.9). The east south central and the midwest pay the highest increase in percentage terms. This is because these regions consume the low-cost, high-sulfur midwestern coal without sulfur restrictions and are forced to substitute high-cost, low-sulfur coal under regulations. Further coal generation accounts for a significant fraction of total generation, so that coal price increases have a proportionally larger effect.

The movement from 4 lb to 1.2 lb SO_2 per million Btu burned is substantially more costly than the movement from 8 to 4 lb. The national average price increases by 4 percent in 1990, 3 percent in 1995 and 2000.

Table 4.8 Total shipments of coal (million tons)

Supply region	Demand region					
	1	2	3	4	5	6
Case: 8 lb SO₂ in 1980						
1	1.88	60.06	110.21	0.07	56.66	38.16
2	0.23	2.49	16.32	0.13	46.74	18.35
3	0	0	55.51	12.90	9.60	31.42
4	0	0	16.56	48.59	0	0.18
5	0	0.01	0.38	0.27	0	0
6	0	0	0	0	0	0
Case: 8 lb SO₂ in 1985						
1	3.00	54.91	29.54	0.05	84.27	0.23
2	0.15	1.66	10.88	0.09	49.16	12.23
3	0	0	145.33	8.60	6.40	82.77
4	0	0	11.04	74.56	0	0.12
5	0	0.01	0.25	0.18	0	0
6	0	0	0	0	0	0
Case: 8 lb SO₂ in 1990						
1	20.77	90.66	14.77	0.02	149.86	0.12
2	0.08	0.83	5.44	0.04	53.67	6.12
3	0	0	215.22	4.30	3.20	117.49
4	0	0	5.52	133.28	0	0.06
5	0	0	0.13	0.09	0	0
6	0	0	0	0	0	0
Case: 8 lb SO₂ in 2000						
1	41.26	72.34	0	0	43.17	0
2	0	0	0	0	8.99	0
3	0	0	195.02	0	183.14	102.40
4	0	0	0	120.98	0	0
5	0	0	0	0	0	0
6	0	0	0	0	0	0
Case: 1.2 lb SO₂ in 1980						
1	1.69	58.33	63.64	0.07	21.00	0.35
2	0.23	2.49	16.32	0.13	79.13	44.36
3	0	0	55.51	12.90	9.60	43.71
4	0	0	79.09	48.60	0	0.18
5	0	0.01	0.38	0.27	0	0
6	0	0	0	0	0	0

7	8	9	10	11	12	Total
0	0	0	0.01	0	0	267.04
3.78	0	0	0	0	0	88.04
0	0	0	1.03	0	0	110.46
0	22.61	0	13.34	28.89	0	130.17
0	3.81	0	0	0	0.01	4.48
19.01	2.81	20.10	0	0	20.33	62.25
0	0	0	0.01	0	0	172.01
2.52	0	0	0	0	0	76.69
0	0	0	0.69	0	0	243.79
0	46.99	0	18.83	53.05	0	204.58
0	2.54	9.11	0	0	0.01	12.10
25.69	1.88	26.35	0	0	34.69	88.61
0	0	0	0	0	0	276.21
1.26	0	0	0	0	0	67.43
0	0	0	0.34	0	0	340.55
0	79.62	0	31.91	87.41	0	337.80
0	1.27	56.43	0	0	0	57.92
31.03	0.94	17.49	0	0	53.13	102.59
0	0	0	0	0	0	156.77
0	0	0	0	0	0	8.99
0	0	0	0	0	0	480.56
0	155.51	106.13	32.09	159.43	0	574.13
40.01	0	0	0	0	0	40.01
0	0	0	0	0	98.96	98.96
0	0	0	0.01	0	0	145.09
3.78	0	0	0	0	0	146.45
0	0	0	1.03	0	0	122.75
0	22.60	0	13.35	28.89	0	192.70
0	3.81	0	0	0	0.01	4.48
18.95	2.81	20.07	0	0	20.32	62.15

Table 4.8 (continued)

Supply region	Demand region					
	1	2	3	4	5	6
Case: 1.2 lb SO$_2$ in 1985						
1	2.27	47.97	29.54	0.05	63.52	58.92
2	0.15	1.66	10.88	0.09	53.33	12.23
3	0	0	42.32	8.60	6.40	18.43
4	0	0	134.92	74.83	0	0.12
5	0	0.01	0.25	0.18	0	0
6	0	0	0	0	0	0
Case: 1.2 lb SO$_2$ in 1990						
1	19.15	81.37	14.77	0.02	140.76	0.12
2	0.08	0.83	5.44	0.04	48.84	6.12
3	0	0	18.50	4.30	3.20	106.31
4	0	0	244.05	135.51	0	0.06
5	0	0	0.13	0.09	0	0
6	0	0	0	0	0	0
Case: 1.2 lb SO$_2$ in 2000						
1	33.63	80.71	0	0	0	0
2	0	0	0	0	30.81	2.35
3	0	0	0	0	183.93	113.40
4	0	0	247.36	130.26	0	5.91
5	1.86	0	0	0	12.26	0
6	0	0	0	0	0	0
Case: 4 lb SO$_2$ in 1980						
1	1.80	59.30	90.01	0.07	34.80	21.83
2	0.23	2.49	16.32	0.13	67.27	18.35
3	0	0	55.51	12.90	9.60	49.14
4	0	0	44.05	48.59	0	0.18
5	0	0.01	0.38	0.27	0	0
6	0	0	0	0	0	0
Case: 4 lb SO$_2$ in 1985						
1	2.77	53.78	85.07	0.05	46.45	5.74
2	0.15	1.66	10.88	0.09	84.91	12.23
3	0	0	37.00	8.60	6.40	75.88
4	0	0	69.39	74.70	0	0.12
5	0	0.01	0.25	0.18	0	0
6	0	0	0	0	0	0

7	8	9	10	11	12	Total
0	0	0	0.01	0	0	202.29
2.52	0	0	0	0	0	80.86
0	0	0	0.69	0	0	76.43
0	46.95	0	18.90	52.98	0	328.71
0	2.54	18.68	0	0	0.01	21.67
25.41	1.88	16.43	0	0	34.71	78.42
0	0	0	0	0	0	256.19
1.26	0	0	0	0	0	62.61
0	0	0	0.34	0	0	132.66
0	79.33	0	32.38	86.36	0	577.69
16.04	1.27	56.78	0	0	7.54	81.85
12.70	0.94	19.86	0	0	44.83	78.33
0	0	0	0	0	0	114.35
0	0	0	0	0	0	33.16
0	0	0	0	0	0	297.33
0	149.32	87.04	34.02	146.91	70.39	871.22
19.81	0	21.75	0	0	0	55.69
22.63	0	0	0	0	35.32	57.94
0	0	0	0.01	0	0	207.83
3.78	0	0	0	0	0	108.56
0	0	0	1.03	0	0	128.18
0	22.62	0	13.34	28.90	0	157.68
0	3.81	0	0	0	0.01	4.48
19.03	2.81	20.10	0	0	20.33	62.26
0	0	0	0.01	0	0	193.86
2.52	0	0	0	0	0	112.44
0	0	0	0.69	0	0	128.58
0	47.07	0	18.87	53.13	0	263.29
0	2.54	9.13	0	0	0.01	12.12
25.77	1.88	26.64	0	0	34.74	89.03

Table 4.8 (continued)

Supply region	Demand region					
	1	2	3	4	5	6
Case: 4 lb SO$_2$ in 1990						
1	20.68	88.20	14.77	0.02	107.06	0.12
2	0.08	0.83	5.44	0.04	94.64	43.60
3	0	0	129.05	4.30	3.20	73.01
4	0	0	110.44	134.01	0	0.06
5	0	0	0.13	0.09	0	0
6	0	0	0	0	0	0
Case: 4 lb SO$_2$ in 2000						
1	40.30	71.18	0	0	116.10	0
2	0	0	0	0	105.37	25.47
3	0	0	115.73	0	0	74.25
4	0	0	99.99	121.92	0	0
5	0	0	0	0	0	0
6	0	0	0	0	0	0

Note: Excludes metallurgical and exports.

The areas that face the steepest increases are the south Atlantic and the east south central regions, followed by the mid-Atlantic and New England regions. Recall that under sulfur regulations these are the areas that turn to scrubbing. The Pacific region suffers the lowest cost increase; it consumes Montana-Wyoming coal both before and after these regulations.

4.3 The Costs of Sulfur Reduction

Sulfur pollution regulations cause increases in the cost of low-sulfur coal, increases in the price of electricity, and a dramatic change in the distribution of coal production and consumption. How much does this cost? There are direct costs in the form of higher prices and indirect costs in the form of higher oil imports and greater strip mining.

Although we consider costs, we have little to say about the benefits of sulfur regulations. The benefit to society of sulfur regulations is a reduction in air pollution, but it is difficult to assess how much that benefit is worth. Scientific evidence is not yet conclusive on the issue of the harm associated with sulfur dioxide emissions.[7] The effect on human health, plant life, and the ecosystem is still subject to controversy.

We cannot hope to settle the issue here. Our goal is much more limited. We ask only how much sulfur reduction costs. Although this is only one-half of the equation, it gives us a standard against which we can compare

7	8	9	10	11	12	Total
0	0	0	0	0	0	230.86
1.26	0	0	0	0	0	145.89
0	0	0	0.34	0	0	209.90
0	80.10	0	32.08	88.00	0	444.70
0	1.27	59.08	0	0	0	60.58
31.22	0.94	15.46	0	0	53.43	105.05
0	0	0	0	0	0	227.58
0	0	0	0	0	0	130.84
20.04	0	0	0	0	0	210.02
0	156.18	107.87	32.28	160.11	0	678.35
20.31	0	0	0	0	0	20.31
0	0	0	0	0	99.36	99.36

the benefits, once they are known. It also gives us an idea of how much of the increase in the cost of electricity is due to cleaning up the environment. Is it a sum worth worrying about?

In addition to estimating costs, we examine the distribution of costs. Some regions gain from sulfur pollution regulations and others lose. We must subtract the gains from the costs to see the net impact. Additionally, and more important, the distribution of gains and losses is at the heart of policy. We must understand that distribution before we can analyze the policy issues.

Who are the losers, and who are the gainers? Consumers must pay higher prices for electricity. Industrial coal users must pay higher prices for coal. Producers of high-sulfur coal lose rents they otherwise might have earned on high-sulfur coal. States that see their coal production diminish lose tax revenues. The gainers are producers of low-sulfur coal, the states that can tax that production and the railroads that will haul that coal. We balance the losers against the gainers to estimate a net cost to society.

The comparison we make here is between no sulfur standard and the current standard. We have already seen that the intermediate standard can be achieved at relatively low cost. The impact of a 4 lb SO_2 standard on the cost of coal is small, and those impacts diminish as they work themselves out through the system—a small effect on coal prices has an

Table 4.9 Price of electricity (mills/kWh in current $)

Case	Demand region			
	1	2	3	4J
1980				
8 lb SO_2 per million Btu	49.04	42.20	37.32	36.86
4 lb SO_2 per million Btu	49.05	42.52	37.59	36.85
1.2 lb SO_2 per million Btu	49.07	43.13	38.24	36.85
1985				
8 lb SO_2 per million Btu	66.59	61.31	50.71	49.58
4 lb SO_2 per million Btu	66.59	61.73	51.39	49.60
1.2 lb SO_2 per million Btu	66.59	65.71	52.50	49.86
1990				
8 lb SO_2 per million Btu	81.96	82.60	65.31	64.36
4 lb SO_2 per million Btu	82.38	83.40	65.82	64.43
1.2 lb SO_2 per million Btu	86.01	87.93	67.51	65.25
1995				
8 lb SO_2 per million Btu	98.87	109.29	87.22	79.68
4 lb SO_2 per million Btu	99.80	109.87	89.43	79.89
1.2 lb SO_2 per million Btu	104.19	114.21	91.22	80.51
2000				
8 lb SO_2 per million Btu	113.12	141.77	115.52	99.92
4 lb SO_2 per million Btu	113.92	142.45	115.00	99.66
1.2 lb SO_2 per million Btu	123.70	144.19	116.82	98.88

Note: The 4J and 8J are aggregates of demand regions 4 and 10 and 8, 11, and 12, respectively.

5	6	7	8J	9	National average
36.61	29.51	40.42	30.97	33.93	36.85
36.94	29.62	40.41	30.98	33.94	37.00
37.58	30.39	40.61	31.04	33.97	37.39
50.07	41.80	48.69	42.63	49.20	50.14
50.20	42.54	48.67	42.66	49.26	50.39
56.04	47.56	49.52	43.11	49.34	52.56
71.73	53.50	64.88	55.45	67.40	66.59
72.11	55.15	65.06	55.45	67.46	67.02
78.45	59.07	66.80	57.31	68.03	69.70
98.01	68.49	84.53	75.43	88.20	87.58
98.00	70.31	84.77	75.63	88.35	87.98
101.83	75.15	87.47	79.44	89.02	90.73
124.42	86.13	115.30	104.84	115.29	113.37
124.62	88.41	115.74	105.20	115.29	113.71
131.58	90.76	115.59	107.84	116.50	116.36

even smaller effect on electricity prices. Given these small impacts for the intermediate standard, our attention focuses on the total costs of meeting the current standards.[8]

The Cost of Electricity to Consumers
Consumers of electricity pay for lower sulfur emission levels in two ways. The direct cost is the increase in the price of electricity. This is directly measurable from the data in tables 4.9, 4.10, and 4.11, shown as area P_1P_2AB in figure 4.1. The second cost is associated with the reduction in electricity consumption. Consumers who would have purchased more electricity at lower prices consume less electricity at the higher price. The value of that lost output is determined by the difference between what

Table 4.10 Demand for electricity (million MWh)

Case	Demand region 1	2	3
1985			
8 lb SO_2 per million Btu	112.7	326.8	524.0
4 lb SO_2 per million Btu	113.1	325.4	520.9
1.2 lb SO_2 per million Btu	113.9	322.2	515.7
1990			
8 lb SO_2 per million Btu	143.9	377.4	631.9
4 lb SO_2 per million Btu	144.4	374.9	625.8
1.2 lb SO_2 per million Btu	144.9	362.0	619.9
1995			
8 lb SO_2 per million Btu	197.6	434.2	735.7
4 lb SO_2 per million Btu	197.7	431.0	731.6
1.2 lb SO_2 per million Btu	194.7	416.1	730.1
2000			
8 lb SO_2 per million Btu	279.0	509.0	841.1
4 lb SO_2 per million Btu	278.4	505.6	839.8
1.2 lb SO_2 per million Btu	272.9	495.1	843.6

Table 4.11 National electricity demand (million MWh)

Case	1975	1980	1985	1990	1995	2000
8 lb SO_2	1,991.868	2,498.4	3,148.5	3,947.3	4,861.1	6,006.4
4 lb SO_2	1,991.868	2,495.0	3,141.1	3,935.4	4,850.4	5,993.2
1.2 lb SO_2	1,991.868	2,488.1	3,110.4	3,844.5	4,746.8	5,922.9

4J	5	6	7	8J	9
260.3	501.1	304.8	382.3	284.7	451.9
260.9	496.5	303.4	383.3	285.1	452.5
262.3	482.2	292.7	381.9	285.4	454.0
344.8	639.1	369.3	473.7	426.5	540.8
346.3	635.0	362.0	475.9	428.6	542.4
350.2	584.3	334.7	473.3	427.1	548.3
446.6	773.6	458.1	568.2	600.1	647.0
449.3	770.8	444.4	571.6	604.1	649.8
459.6	720.8	415.1	565.7	583.7	661.1
586.6	994.7	577.7	662.9	752.0	803.5
590.5	994.3	555.6	666.7	755.3	807.1
613.6	959.3	527.4	667.8	717.8	825.3

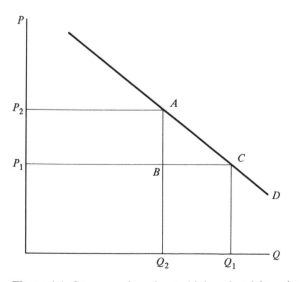

Figure 4.1 Consumer loss due to higher electricity prices

consumers would have paid, the value to them, and the cost to them of the electricity. This is shown as area *ABC* in figure 4.1. We assume linearity of the demand function. The range over which demand varies is small, so this assumption will not significantly bias the results.

The costs are calculated for three representative years: 1985, 1990, and 2000. We assume the full force of sulfur pollution regulations in the first year, 1985. The year 2000 allows nuclear power some response to tighter sulfur pollution restrictions. The year 1990 is included to give a picture of medium-term effects.

Costs are also calculated on a regional basis in order to understand the distributional implications. The results are displayed in table 4.12.

Small percentage changes in the cost of electricity correspond to large absolute amounts of money because of the enormous amount of electricity consumed each year. One mill per kWh is roughly 3 percent of the national average price of electricity. Since approximately 2,000 billion kilowatt-hours were consumed in 1977, a one-mill increase in the cost of electricity translates into $2 billion in direct costs alone.

The 1.2 lb SO_2 standard costs over $5 billion (in 1979 dollars) per year in 1985. The costs rise to almost $7 billion dollars per year in 1990 and fall to roughly $6 billion in 2000. This represents consumer surplus lost.

However, focusing on the national impact obscures the highly skewed regional distribution of cost impacts. These antipollution policies will have their greatest impact on air quality in the eastern states, where high-sulfur coal would be burned in the absence of sulfur regulations. It is also in the eastern region that the greatest cost is paid for lowering air pollution. The estimated cost in the east for 1.2 lb SO_2 ranges from $327 million per year (in 1979 dollars) in 1990 in New England to over $2.2 billion in 1990 in the south Atlantic.

These numbers have to be adjusted for the size of the various areas. Table 4.13 shows the cost per capita. We use the census bureau's forecast of 1990 population by region to calculate these figures.[9] On a per-capita basis the cost ranges from $5.80 in the Pacific to $70.00 in the east south central region. The relatively high cost in region 8 is due to the effect on

Table 4.12 Cost of alternative sulfur standards to consumers of electricity (billion 1979 $)

Case	1985	1990	2000
4 lb SO_2	1.1	0.9	0.8
1.2 lb SO_2	5.4	6.8	5.8

the southern extremity of that region. In sum air pollution regulations cost the most where high-sulfur coal is the cheapest source of coal.[10]

The Cost to Industrial Consumers of Coal

We have already seen that industrial users of coal pay much higher prices under sulfur regulations. We have assumed they consume low-sulfur coal. If this were not the case, they would have to scrub to meet pollution regulations, and the costs of using coal would be much higher. The model results for the electric utility sector validate this assumption with respect to industrial coal use. Under sulfur regulations, only the east chooses to scrub. Furthermore, there are economies of scale in scrubbing, so industrial facilities smaller than electric utilities would face even less favorable circumstances under scrubbing.

The cost to industrial consumers consists of the consumers' surplus lost. Part of this loss is simply the difference in coal price multiplied by their new level of consumption. The other part is the lost opportunity to purchase coal at prices below what they would be willing to pay. However, this is likely to be quite small, since the demand response to the higher prices in absolute terms is small.

The calculation of consumer surplus lost should be performed on a region-by-region basis. Table 4.14 presents the results of the calculation for the year 1990. The bulk of the cost is paid in region 5. The total cost is $80 million per year in 1990, equivalent to $44 million per year (in 1979 dollars). This is a small loss compared to the loss borne by electricity consumers (including the industrial electric consumers).

Gainers and Losers, Producers of Coal

Producers of coal no longer in demand lose rents. This means that the high-sulfur coal producers of northern Appalachia and the midwest lose as tightening sulfur regulations lead to a decline in the demand for their coal. It is not that the supply function is shifting in any way but that we are moving up and down individual supply curves. We are able to calculate the loss in producers' surplus by comparing production and prices under each scenario on each of eight supply functions for each of the six supply regions. Table 4.15 shows the result of this exercise for the year 1990. The eastern predominantly high-sulfur coal producers lose a total of $1175.7 million per year in 1990. This is a loss of $653 million (in 1979 dollars). The western regions and eastern low-sulfur producers gain. Here the net effect of sulfur pollution regulations on coal producers is a

Table 4.13 Cost to electric consumers of alternative sulfur levels by region in 1990 (million 1979 $)

	Demand region		
Case	1	2	3
4 lb SO$_2$	33.7	167.0	178.0
1.2 lb SO$_2$	326.7	1093.4	764.1
Cost per capita of 1.2 lb SO$_2$	24.0	52.6	17.6

Table 4.14 Loss to industrial consumers of a 1.2 lb SO$_2$ sulfur standard in 1980 (million 1979 $)

Demand region	Increase in coal price (¢/million Btu)	Reduction in consumption (10^6 Btu)		Cost
1	49.4	3.4×10^6		1.7
2	49.2	1.2×10^7		5.9
3	15.8	5.8×10^6		0.8
4J	5.2	2.4×10^6		0.1
5	40.1	7.7×10^7		30.8
6	50.1	5.6×10^6		2.8
7	12.6	1.4×10^7		1.8
8J	11.2	3.0×10^6		0.3
9	3.4	2.4×10^5		neg.
			Total	44.2

gain of only $14 million per year (in 1979 dollars). This is pure rent and amounts to $.01 per ton of coal, a trivial amount.

The net change is certainly less than the error in measurement. However, the net change conceals substantial interregional redistribution. The high-sulfur regions of northern Appalachia and the midwest lose to the low-sulfur regions of southern Appalachia and the western producing regions. Within each region low-sulfur producers gain at the expense of high-sulfur coal producers. In addition the results show how sulfur content and Btu value interact in determining the value of low-sulfur coal. The point at which the change in rents becomes negative is at a lower sulfur category in the western regions, where Btu values are lower. This occurs because, as heating values decline, the coal must be lower in sulfur by weight in order to meet the emissions standards.

In sum substantial redistribution occurs, but it is almost pure transfer between various segments of the coal industry. This accounts for the highly charged political controversy that accompanies any environmental

4J	5	6	7	8J	9
13.5	134.3	334.8	47.7	0	18.0
172.9	2181.0	1088.0	504.5	440.8	193.0
9.5	53.4	70.0	19.9	34.1	5.8

policy action, even though the net results are small. This controversy creates uncertainty with respect to policy, and this uncertainty is potentially more costly than the policy itself.

The Effect of Sulfur Regulation on Tax Revenues

Severance taxes in the coal industry have been increasing. The levels of taxes now range from 0 to over 30 percent.[11] Generally the western states have higher taxes, reflecting their favorable competitive position. The increased demand for their low-sulfur coal has increased their ability to raise taxes.

Since the coal industry itself is competitive, it is difficult for coal producers to exercise monopoly power. The rents discussed earlier in connection with the coal industry arise in a competitive natural-resource industry. State legislatures are, however, in a position to affect the price of coal. The legislatures can set taxes that capture rents for themselves. In fact, functioning as monopolists, they can raise the level of their rents. The tax in the west actually depends upon market conditions, upon the demand for low-sulfur coal and the supply of eastern low-sulfur coal. The western states, particularly Montana and Wyoming, can function as monopolists and set a tax that maximizes their revenues. Research reported elsewhere indicates that this is in fact occurring.[12] Here we assume that tax levels are fixed at current levels. The goal is to analyze how changes in sulfur regulations, holding tax levels constant, affect costs and the distribution of rent.

Table 4.16 shows the impact of sulfur regulations on tax revenues. In 1990 tax revenues to coal-producing states increase by 142 percent from $2,226 to $3,170 million. The net increase is about $944 million in 1990, or about $525 million (in 1979 dollars). The bulk of the gain is in region 4, and the bulk of the loss is in region 3. Again, large regional changes cancel each other. However, the net result is still a substantial gain to coal states as a whole.

Table 4.15 Change in rents to coal producers by region and sulfur content in 1990 due to 1.2 lb SO$_2$ standard (million 1979 $)

Supply region	Sulfur category			
	1	2	3	4
1	3.5	31.5	0.4	0
2	21.4	58.3	41.9	0.4
3	0	0	8.3	7.9
4	323.0	3.3	−0.1	−9.7
5	107.1	−1.7	−5.6	−2.1
6	264.7	−27.6	−29.8	−2.9

[a]Total does not add to sum of individual regions due to independent rounding.

Table 4.16 Tax revenues (million current $)

Case	Supply region						Total
	1	2	3	4	5	6	
1980							
8 lb SO$_2$	95.5	225.5	64.8	123.4	3.77	31.4	544.4
4 lb SO$_2$	75.4	247.0	88.1	151.6	3.77	31.4	597.3
1.2 lb SO$_2$	68.2	290.3	83.8	188.7	3.77	31.7	666.5
1985							
8 lb SO$_2$	118.5	341.1	365.7	291.4	10.7	66.2	1,193.6
4 lb SO$_2$	130.8	399.6	181.8	379.7	10.8	66.5	1,169.1
1.2 lb SO$_2$	117.3	348.6	109.9	519.4	18.0	63.9	1,177.0
1990							
8 lb SO$_2$	232.4	468.7	704.8	660.1	50.1	109.9	2,226.0
4 lb SO$_2$	213.2	674.4	425.6	884.5	52.1	108.3	2,353.2
1.2 lb SO$_2$	213.2	473.7	270.1	2,029.8	79.8	103.9	3,170.5
2000							
8 lb SO$_2$	342.0	868.5	1,864.7	3,148.1	83.3	267.2	6,573.8
4 lb SO$_2$	468.0	1,430.1	772.5	3,869.7	56.8	272.2	6,869.3
1.2 lb SO$_2$	334.5	982.0	1,097.5	5,879.4	126.7	110.7	8,530.8
Total (present value in 1977 dollars)							
8 lb SO$_2$	798.6	2,105.4	2,020.1	2,153.9	99.0	392.7	7,569.6
4 lb SO$_2$	782.0	2,574.5	1,241.7	2,867.4	104.1	393.5	7,963.2
1.2 lb SO$_2$	728.6	2,452.3	922.1	5,157.3	162.1	306.4	9,728.8

Note: Discount rate is 10 percent real.

5	6	7	8	Total[a]
0	0	− 176.0	− 358.0	499.4
− 43.1	− 14.8	− 2.3	− 6.8	55.0
0.5	0	0	− 169.7	− 153.0
− 2.2	− 0.3	− 0.2	− 0.3	313.4
− 1.1	− 0.8	− 0.5	− 0.9	94.4
− 0.3	− 0.1	0	0	− 204.0

Net Direct Costs of Sulfur Regulations

Now we are ready to total up gains and losses to the United States. Recall that the most important benefit—clean air—is omitted. The value of our estimate is simply that it provides a benchmark comparison.

The total loss figure is dominated by the loss to consumers of electricity. In 1990 they lose $7 billion per year (in 1979 dollars). The loss to industrial consumers is small, amounting to $.044 billion. The total losses are $7.431 billion. From that we subtract over one-half billion dollars per year in increased taxes, for a net loss of $6.89 billion per year in 1990 (in 1979 dollars). This is the estimated total direct cost of sulfur pollution regulation.

The cost calculation is overwhelmingly dominated by electricity cost increases. Table 4.12 shows that this cost is fairly constant throughout the period, declining slightly by the year 2000. Thus the total-cost measure for 1990 is approximately the annual cost throughout the period. The outlay is large. Seven billion dollars per year is equivalent to 10 percent of our current oil import bill and half of our current national coal bill.[13] These costs, however, must be weighed against the benefits of reduced air pollution.[14] The definitive balancing of these costs and benefits must await more information.

Indirect Costs of Sulfur Regulations

In addition to direct costs there are several nonquantifiable costs. Western output of strip-mined coal increases. The true environmental cost of this phenomenon is not known. We have included strip-mine reclamation costs in our cost estimates but no one knows the true cost of strip-mining in the west. In fact the true cost is incalculable because aesthetics and social values are involved. The best we can do is say that the cost includes an increase of 50 percent in the output of western coal along with the annual $7 billion of directly estimated costs. The second indirect cost is the effect on national energy independence. Oil consumption rises as coal consumption declines. What effect does sulfur pollution regulation have on oil imports?

4.4 Sulfur Regulations and Oil Imports

The reduction of oil imports has been a goal of each of the last three administrations.[15] Recently the viewpoint was advanced that we might have to relax environmental goals and loosen standards in order to increase the use of coal and decrease oil imports. The results of these scenarios suggest that the effect of environmental standards on industrial oil consumption is small. Table 4.17 shows the impact of sulfur standards on oil consumption in the industrial sector. Pollution standards result in an increase in oil consumption of only 20,000 barrels per day in 1985 and 62,000 barrels per day in 2000, far less than the decrease in coal consumption. In the year

Table 4.17 Industrial use of gas, oil, and electricity (10^{16} Btu)

Fuel	1985	1990	1995	2000
8 lb SO$_2$				
Oil	0.1711	0.1685	0.1621	0.1499
Gas	0.5826	0.4282	0.3368	0.2868
Electricity	0.5717	0.7460	0.9264	1.117
4 lb SO$_2$				
Oil	0.1723	0.1705	0.1641	0.1517
Gas	0.5863	0.4330	0.3410	0.2901
Electricity	0.5709	0.7449	0.9260	1.117
1.2 lb SO$_2$				
Oil	0.1757	0.1798	0.1767	0.1635
Gas	0.5974	0.4565	0.3671	0.3127
Electricity	0.5662	0.7314	0.9131	1.114
Use of oil (10^{16} Btu)				
8 lb SO$_2$	0.1235	0.2006	0.0809	0.0658
4 lb SO$_2$	0.1235	0.2016	0.0855	0.0659
1.2 lb SO$_2$	0.1246	0.2106	0.1281	0.1242

2000 the reduction in coal consumption is 25.2 million tons of coal of 22 million Btu per ton. This is equivalent to 253,000 barrels per day, far more than the estimated increase in oil consumption. The difference between the decline in the oil-equivalent consumption of coal and the increase in oil consumption is not filled by gas or by electricity (see table 4.17) but is rather due to a reduction in the demand for energy by the industrial sector (see table 4.18). The estimates here are subject to great uncertainty. Nevertheless, the qualitative message is important. Loosening sulfur pollution regulations would gain very little in terms of import reduction in the industrial sector.

Utility consumption of oil is affected somewhat more by the imposition of sulfur standards. Sulfur standards, given our relatively low oil prices, account for an increase in oil consumption of 266,000 barrels per day in the utility sector by 2000 (table 4.17). In earlier years the difference is quite a bit smaller. Thus the maximum yearly increase in total oil consumption is about 330,000 barrels per day. This is not a dramatic increase when one considers it corresponds to the difference between no sulfur standards and the current 1.2 lb standards. *Only an extremely high premium on import reduction would justify abandoning sulfur regulations.*[16]

4.5 Reducing Western Mining

Sulfur reduction entails substantial expansion of western output. Western output expands by 50 percent under the 1.2 lb SO_2 standard. This expansion entails environmental costs we cannot measure. How costly is it to reverse the movement to the west, given sulfur regulations of 1.2 lb SO_2 and the factor price developments considered earlier? Answering this question will give us a standard against which to compare real or imagined costs of western output. We will analyze several methods for reducing output, each of which represents an area of actual policy interest.

The motivations for these policies range from environmental concern to desires to capture rents. For example, some groups fear the disruptive effects of boomtowns, whose social and psychological costs motivate

Table 4.18 Industrial demand for energy (10^{15} Btu)

Case	1980	1985	1990	1995	2000
8 lb SO_2	15.3	15.7	16.3	17.7	19.4
4 lb SO_2	15.3	15.7	16.3	17.6	19.4
1.2 lb SO_2	15.3	15.6	16.2	17.5	19.2

their desire to curtail western output. There are those who are concerned with damage to the land and are not satisfied that adequate reclamation is possible. The concerns of these groups are environmental externalities. They feel there are costs not reflected in the current price of coal. Finally, as pointed out earlier, there are those who seek to raise the price of western coal in order to increase their rents. The latter is a distributional goal that entails efficiency loss.

Regardless of motivation the effect of these various policies is to reduce the output of western coal. However, the distinction between goals is relevant in assessing policy. If one believes there are strong externalities associated with western production, the reduction of western output is efficient policy. If, on the other hand, one attributes a small cost to the environmental externalities, then the cutback in western output and the associated higher costs are inefficient policy.[17]

Our strategy is not to try to mimic policy exactly but rather to bracket outcomes. Policies toward leasing of western coal provide an example. There is currently a moratorium on leasing federal western coal lands. We do not attempt to guess when this moratorium will be relaxed.[18] We ask a simpler question, and one more meaningful for an analysis of policy: What would happen if such a moratorium were continued indefinitely? In that way we can assess the impact of leasing western coal on the evolution of the coal industry. A result indicating high costs would be evidence of the importance of leasing policy. In this way we can highlight important policy areas and bracket the costs involved. Where the cost of a policy action is small, even under extreme assumptions, we can confidently rule it out as likely to affect coal development and/or the electric utility industry.

Best Available Control Technology (BACT)

The policy subject to the greatest controversy is the use of the Best Available Control Technology (BACT) for sulfur removal. This policy debate revolves around how to implement the 1978 amendments to the Clean Air Act. These amendments called for the use of "best available control technology" to reduce sulfur emissions. It is not enough to burn low-sulfur coal. Rather some mechanical or chemical device must be used to reduce sulfur emissions.

The exact formulation of these regulations was left to the administrator of the EPA. Preliminary standards were promulgated in September 1978 and revised in June 1979. The preliminary standards called for a 90 percent reduction in sulfur emissions regardless of the sulfur content of the coal. This, in effect, means that all coal has to be scrubbed since stack gas scrubbing devices are the only technology capable of achieving such a level of sulfur reduction. The final standards were modified to allow for more flexible standards. Now scrubbing at a 90 percent level is necessary only when emissions exceed 0.6 lb SO_2; below that level scrubbing is not necessary. The minimum level of scrubbing, regardless of emissions, is 70 percent sulfur removal.

The motivations behind this policy reflect both efficiency and distributional goals. The effect of scrubbing regulations is to reduce western output. The logic is simple. If one has to scrub both low- and high-sulfur coal, it does not make economic sense to pay a premium for low-sulfur coal. Rather than pay for western coal, eastern utilities would rather scrub high-sulfur coal. The point here is that BACT has very important distributional implications. And these distributional implications lie behind a great deal of the motivation for BACT. Eastern coal interests supported this policy. The preliminary standards were modified in response to pressure from western interests. The so-called sliding scale of BACT meant that western markets would be protected to some extent since scrubbing low-sulfur coal to 70 percent effectiveness is cheaper than scrubbing high-sulfur coal to 90 percent.

Distributional goals, however, are not the sole motivation behind this policy. Environmental goals are also served. In the first place western coal output—thus strip mining—is reduced. Second, sulfur emissions will be reduced. In areas where low-sulfur coal would be burned anyway, forced scrubbing lowers total emissions. This is counterbalanced to a small extent by areas that switch to scrubbing high-sulfur coal. It is possible that emissions from scrubbing high-sulfur coal are higher than emissions had the low-sulfur coal been burned instead. But net emissions for the United States will certainly be reduced.

Opponents of BACT claim that there are more efficient ways to reduce sulfur emissions. Setting standards and allowing them to be met in any feasible way, including use of low-sulfur coal, would be cheaper. They also argue that scrubbing itself creates an environmental problem of solid waste disposal. In the stack gas scrubber the sulfur in effluent gases combines with limestone to produce calcium sulfate, a solid with the consistency of paste. Removal of this waste is a costly process.[19] Finally, opponents claim that the devices themselves are unreliable and will not be able to achieve the efficiency levels mandated by regulations. In sum the opponents argue on efficiency grounds that address only part of the motivations behind these regulations.

The Effects of BACT

Our model has implemented the extreme version of BACT. In keeping with the desire to bracket outcomes, we have assumed that after 1983 all new coal plants must use stack gas scrubbing devices.[20] This was implemented in the model by eliminating the choice of a coal plant without scrubbers. As expected, the effect of BACT is to change the regional allocation of output in favor of the eastern producing regions.

We compare the scenario of tight sulfur standards to a scenario with tight sulfur standards plus BACT. Table 4.19 shows the effect on regional output (compare this to table 4.7, 1.2 lb SO_2 case). Because of the long lead time involved in implementing this policy, the effect is not important until the 1995 to 2000 period. In 1995 Montana-Wyoming output is reduced by almost 30 percent or 185 million tons per year. In 2000 the relative decline in western output is almost 100 million tons.

While Montana-Wyoming output declines by 100 million tons, eastern output increases only by an estimated 29 millions tons in 2000. A large part of the lost Montana-Wyoming output is made up by increases in Utah-Colorado and in Arizona-New Mexico output. Under BACT the latter region is able to expand its production of high-sulfur coal. The surprising result is that these two western regions gain under BACT. The changes for the east are important in the middle part of our period, but by the year 2000 the net change is small.

We can understand this result more easily by examining distributional patterns. Tables 4.20 and 4.21 show the distribution of coal in the years 1995 and 2000 with BACT. The big change under BACT, as can be seen by comparing table 4.20 to table 4.8, is that the midwestern consuming region uses coal from the midwestern producing region. This is the swing region. The rest of the eastern consuming regions continue to consume eastern coal with or without BACT. The effect in the midwest diminishes between 1995 and 2000 because total coal demand declines as new nuclear power plants come on-stream (although the addition of these nuclear

Table 4.19 Production by region under 1.2 lb SO_2 sulfur standard with BACT (million tons)

Year	Region					
	1	2	3	4	5	6
1980	201.1	251.5	129.8	192.7	13.5	62.1
1985	183.0	288.7	84.4	357.4	25.8	87.5
1990	322.8	196.2	178.7	559.3	63.4	105.0
2000	251.6	172.7	300.5	761.4	74.3	90.6

power plants is at best problematic). The result here should be taken to indicate only that BACT exerts its primary influence in the midwest region, and, depending upon the role of nuclear power, its effect can be substantial.

The increase in Arizona-New Mexico production comes about due to the switch in demand region 12 (Arizona-New Mexico) from using Powder River Basin coal to coal from its own coalfields. In essence BACT keeps coal consumers close to home in their search for coal. It is no longer economic to transport low-sulfur coal over long distances. The greatest loser is the Powder River Basin, since that area must ship its coal the farthest.

BACT increases the demand for high-sulfur coal and lowers the demand for low-sulfur coal, thereby narrowing the sulfur premium. Table 4.22 shows the net changes in delivered prices. The differentials narrow, but the extent of the narrowing is small. The low-sulfur coal price in 2000 is in general only about 2 percent lower, and the high-sulfur coal price is on the order of 1 percent higher. Recall that these are delivered prices, so the effect of any change in mine-mouth prices is diluted by transport costs.

It is not surprising that the effect on the differential is small. We have already seen that the main effect of BACT is to switch from low-sulfur western coal to high-sulfur eastern coal in the midwest consuming region. That is a shift from one elastic supply curve to another. Under these circumstances, relatively large output changes among regions occur with relatively small price effects.

The cost of BACT is therefore not in higher coal prices, but rather in the cost of using a more expensive technology for generating electricity. Scrubbers must now be used in regions where economics alone would not have dictated that choice.

The effect of BACT is to raise electricity prices. The national average price increases by 1.4 percent in 1990, 1.7 percent in 1995, and 0.5 percent in the year 2000 (see table 4.23). These are increases above the cost of sulfur standards. These percentage increases are small, but not surprising. The percentage of total electric generation in the year 2000 accounted for by coal is on the order of one-third of total generation. Scrubbing costs are on the order of 10 percent of generation costs. Generation cost is about 50 percent of total delivered costs of electricity. Thus BACT could be expected to increase costs on the order of 1 to 2 percent ($0.33 \times 0.1 \times 0.5 = 0.0165$). However, plants with scrubbers can use a cheaper fuel,

Table 4.20 Coal distribution in 1995 with BACT and 1.2 lb SO$_2$ sulfur standard (million tons)

Supply region	Total shipments in demand region					
	1	2	3	4	5	6
1	27.18	83.79	0	0	140.13	0
2	0	0	0	0	73.16	2.31
3	0	0	85.85	0	0	95.03
4	0	0	216.74	141.68	0	5.80
5	0	0	0	0	5.88	0
6	0	0	0	0	0	0

Note: Excludes metallurgical and export.

Table 4.21 Coal distribution in 2000 with BACT and 1.2 lb SO$_2$ sulfur standard (million tons)

Supply region	Demand region					
	1	2	3	4	5	6
1	33.47	76.64	0	0	57.46	0
2	0	0	0	0	13.23	2.43
3	0	0	47.17	0	119.66	123.63
4	0	0	189.69	117.12	0	6.11
5	1.88	5.01	0	0	6.55	0
6	0	0	0	0	0	0

Note: Excludes metallurgical and export.

and thus 1 to 2 percent is an upper limit for the increase in average electricity price.

This relatively small effect, however, amounts to about $3 billion (in 1979 dollars) per year in 1995. Under BACT total electric demand is 4,677.4 × 10^9 kWh, down from 4,746.8 × 10^9 kWh without BACT in 1995. In 2000 the cost is over $1 billion (in 1979 dollars) per year. The regional impacts are highly uneven. Some regions have only a small rise or even a lowering of their electricity prices while others experience increases exceeding 6 percent.[21] This regional impact, together with regional production effects, accounts for the highly controversial nature of BACT.

The western regions pay the highest price. These are regions that would consume western coal without BACT regulations. They continue consuming western coal after imposition of these regulations. The only difference is that they must scrub the low-sulfur coal. Thus regions 4, 7, 8, and 9 pay

7	8	9	10	11	12	Total
0	0	0	0	0	0	251.11
0	0	0	0	0	0	75.47
0	0	0	0	0	0	180.88
0	116.73	27.35	35.34	119.60	0	663.26
16.02	0	56.81	0	0	5.27	83.98
24.16	0	0	0	0	68.94	93.09

7	8	9	10	11	12	Total
0	0	0	0	0	0	167.57
0	0	0	0	0	0	15.66
0	0	0	0	0	0	290.46
0	159.02	88.44	31.91	155.80	13.31	761.40
41.24	0	6.66	0	0	0	61.34
0	0	0	0	0	90.58	90.58

the bulk of the cost. The eastern regions—1, 2, 5, and 6—pay no cost (recall that these regions choose to scrub by 1995 even without BACT). Again, the swing status of region 3 emerges. This midwestern consuming region switches a significant amount of its lower-sulfur coal consumption to scrubbing high-sulfur coal. The cost impact is moderated somewhat since high-sulfur coal is a cheaper fuel. Thus the impact in the midwest lies between the zero impact in the east and the large impact in the west.

The BACT regulation costs consumers $3 billion per year in 1995 and declines thereafter. If viewed as a policy to protect eastern coal markets, its benefit is the additional cumulative production of 900 million tons of coal in the east over the 1975 to 2000 period (see table 4.24).

If BACT is viewed purely as an environmental policy, the benefits of reduced pollution must be added. But the most efficient way to reduce pollution is to tax sulfur emissions. A less efficient alternative, but still more efficient than BACT, is to specify a lower emission standard, then let utilities decide how best to meet the regulation. In fact BACT is

Table 4.22 The effect of BACT on the delivered cost of low-sulfur coal (¢/million Btu in current $)

	Demand region			
Case	1	2	3	4
1980				
Without BACT	188.8	175.8	157.3	118.5
With BACT	188.8	175.8	157.3	118.5
1985				
Without BACT	311.9	293.2	240.3	175.9
With BACT	311.8	293.2	240.0	175.5
1990				
Without BACT	419.6	395.3	314.3	230.3
With BACT	419.6	395.3	313.6	228.9
2000				
Without BACT	742.2	700.8	520.0	383.1
With BACT	730.1	693.8	507.8	369.1

Table 4.23 Electricity prices (mills/kWh in current $)

	1990		1995		2000	
Demand region	With BACT	Without BACT	With BACT	Without BACT	With BACT	Without BACT
1	85.8	86.0	104.0	104.2	125.1	123.7
2	87.8	87.9	114.5	114.2	143.8	144.2
3	66.8	67.5	92.1	91.2	117.7	116.8
4J	69.2	65.2	85.3	80.5	103.0	98.9
5	77.6	78.5	101.3	101.8	125.8	131.6
6	60.7	59.1	74.9	75.2	91.5	90.8
7	65.9	66.8	89.6	87.5	121.6	115.6
8J	62.2	57.3	81.6	79.4	104.5	107.8
9	69.9	68.0	91.7	89.0	119.9	116.5
National average	70.7	69.7	92.29	90.7	117.0	116.4

5	6	7	8	9
180.5	159.3	145.8	82.9	141.2
180.5	159.3	145.8	82.9	141.2
303.7	278.6	224.1	123.5	208.0
304.1	279.1	221.3	122.7	207.3
411.1	376.4	308.2	179.6	282.9
411.4	376.9	303.4	167.9	277.9
710.8	637.3	541.8	314.7	485.1
694.4	626.8	538.7	323.0	474.6

roughly the equivalent of a standard of 0.6 lb SO_2. At this level, even western coal would need to be scrubbed.[22]

The east-west totals conceal the very different regional effects of BACT. Montana-Wyoming cumulative output decreases by 1.3 billion tons. The midwest gains by over 600 million tons. The surprise is the gain of approximately 630 million tons realized by Arizona-New Mexico.

In sum BACT appears most effective as a means of reducing air pollution and should be viewed in those terms. The cost of $3 billion could be lessened somewhat by allowing less than 70 percent scrubbing, but the change is likely to be small. Again, the interregional transfers are large, and that explains why BACT has been so controversial.

Leasing Policy
The federal government owns mineral rights to vast quantities of western low-sulfur coal (see table 4.25), but since 1971 there has been a moratorium on leases of these federal coal lands. Leasing is now scheduled to begin in 1980 or 1981.

The reasons behind the long-lived moratorium are environmental. The original moratorium was instituted to give the federal government time to consider the economic implications of its leasing policies, particularly the manner in which leases were sold to private developers. It was felt that

Table 4.24 Cumulative output, 1975 to 2000 (million tons)

	Supply region		
Case	1	2	3
1.2 lb SO$_2$ without BACT	6,154.9	5,167.8	3,800.0
1.2 lb SO$_2$ with BACT	6,317.9	5,253.3	4,472.7

Table 4.25 Federal ownership of western lands

State	Percent federal land	Percent federal land currently under lease	Percent total land available under moratorium
Montana	41	2.7	20.2
Wyoming	50	3.6	35.5
Colorado	53	2.0	21.5
Utah	66	61.5	74.6
Arizona	0	n.a.	100.0
New Mexico	40	18.9	35.2

Source: Federal Coal Leasing Amendments Act of 1975, P.L. 94–377, *Legislative History,* p. 1945.

leases were being bought by speculators who resold the land rather than develop it. Attempts by the government to resume leasing, however, have been held up by environmental challenges. The most important of these was a suit brought by the Sierra Club, which has been settled, that sought to require an areawide environmental impact statement.

The impact of a leasing moratorium is twofold. In the first place government land cannot be mined. Second, because of the checkerboard pattern of ownership and economies of scale in mining, the ineligibility of government reserve parcels for mining implies the ineligibility of an equal amount of private land.[23] We use the leasing moratorium as a way to analyze the importance of western coal.[24]

The leasing moratorium is simulated in the following way. Data are available on the percentage of federally owned coal lands in each state that are not now under lease. We assume that keeping a government-owned parcel from being mined implies that a private parcel also would not be mined. The reserve base in the state was then reduced by the amount of coal contained in ineligible reserves. We are assuming that the federal land not yet leased and private land not yet under development are

4	5	6	Total east	Total west
12,306.4	1,392.8	1,432.0	15,432.7	15,130.8
10,966.5	1,275.8	2,061.8	16,043.9	14,304.1
		Difference	+911.2	−826.7

distributed independently of overburden ratios and other cost-determining factors (see table 4.25). The simulation was then equivalent in all other respects to the BACT scenario. Any further reduction in western output is due to the incremental addition of a leasing moratorium. The government owns such a large fraction of coal reserves that an indefinite moratorium severely restricts western production in the latter part of the period. Although this policy will not be continued, our analysis allows us to examine the importance of western coal.

Table 4.26 shows the impact of a leasing moratorium on production. Comparing the production results to the previous case (table 4.19), it is clear that the main impact of a leasing moratorium is not felt until the end of the period. In 1995 the major effect is a shift from the Arizona-New Mexico to the Utah-Colorado producing region. The Montana-Wyoming region is affected only to a small degree. By 1990 the effect on Montana and Wyoming is large, amounting to a reduction of over 140 million tons per year. This loss is not filled by eastern output in 1990, because the reduction in western output results from a reduction in the total demand for coal. This of course results from price effects.

After 1990 there is both a reduction in western output and an increase in eastern output, particularly in the midwest. By 2000 the effects are very large. Eastern output is up by 46 percent, and western output is down by almost 48 percent. The leasing moratorium creates a need for very high levels of production in the midwest. This essentially transfers western environmental problems to the midwest. Although the midwest expands substantially by underground mining, its ability to expand to that level is questionable.[25] Surely some of the same reactions against mining now being experienced in the west will be voiced in the midwest as the industry tries to attain the higher production seen here.

The effect of holding back the only large incremental source of low-sulfur coal is to substantially raise the cost of low-sulfur coal (see table 4.27). The largest impact is felt in the western consuming regions, since they turn for all their coal to the western producing areas. The eastern regions, relying on the east for both high- and low-sulfur coal, are relatively unaffected. The most important effect, however, is the increase in the delivered price of high-sulfur coal in the west.[26]

This high-sulfur coal price increase is important because it has a serious impact on electricity prices. BACT causes an increase in the amount of high-sulfur coal used. Table 4.28 shows the coal consumed (in Btu) by low- and high-sulfur categories for the years 1990, 1995, and 2000. By the end of the period no low-sulfur coal is used by the utility sector in the eastern regions 1, 2, 5, and 6. Low-sulfur coal is used in significant amounts in the utility sector only in the midwest and in the southwest. Elsewhere high-sulfur coal is used because of scrubbing requirements. Thus leasing requirements have the greatest effect on electricity prices by raising the cost of high-sulfur coal in the west.

Table 4.26 Production by region under a leasing moratorium (million tons)

	Region					
Year	1	2	3	4	5	6
1980	201.1	251.5	129.8	192.6	21.0	53.7
1985	198.0	280.2	85.9	344.7	50.8	57.6
1990	332.6	190.6	179.5	418.0	44.9	62.3
2000	263.4	191.1	488.2	386.6	110.0	2.6

Table 4.27 Delivered price of coal under a leasing moratorium (current $/million Btu)

	Demand region			
Case	1	2	3	4J
1980				
High sulfur	1.44	1.33	1.32	1.06
Low sulfur	1.71	1.59	1.45	1.10
1985				
High sulfur	2.11	1.94	1.93	1.59
Low sulfur	2.82	2.64	2.26	1.66
1990				
High sulfur	2.94	2.72	2.64	2.17
Low sulfur	3.96	3.73	3.10	2.29
2000				
High sulfur	5.58	5.19	4.65	4.52
Low sulfur	7.50	7.11	6.28	4.91

The impact on electricity prices presents a familiar picture (see table 4.29). The national impact appears small, but impacts in some individual regions are large. Leasing hurts the western and midwestern regions, leading to as much as a 4.2 percent increase in midwestern electricity prices in 2000.

A factor that provides a safety valve even under a leasing moratorium is an expansion of nuclear power. By the year 2000 an additional 28 gigawatts of nuclear capacity are built in the leasing moratorium scenario (not shown). This results from the higher price of coal and serves to moderate the effect of a restricted leasing policy on electricity prices.

The electric utility industry is able to insulate itself to a large degree from a leasing moratorium because of its ability to substitute eastern high-sulfur for western low-sulfur coal and nuclear power for coal power. Those consumers who must consume low-sulfur coal will of course face much higher prices for coal and can be expected to consume less. (Consequently there is a substantial reduction in coal use in the industrial sector, see table 4.30.)

This analysis of leasing policy has contributed several insights into the importance of western coal. Severely restricting the development of western coal affects the price of high-sulfur coal as well as that of low-sulfur coal, with the largest impact in the western consuming regions.

5	6	7	8 J	9
1.40	1.24	1.31	.77	1.28
1.63	1.43	1.38	.79	1.32
2.16	1.90	1.93	1.06	1.87
2.75	2.49	2.21	1.26	2.02
3.02	2.60	2.81	1.51	2.74
3.88	3.55	3.15	1.90	2.89
5.71	4.56	5.38	3.93	5.54
7.30	6.72	6.40	4.41	5.75

Table 4.28 Utility coal consumption by sulfur content under a coal-leasing moratorium (million tons of 22 million Btu/ton)

	Demand region			
Case	1	2	3	4*J*
1980				
Low sulfur	.495	50.4	172.4	60.2
High sulfur	0	0	0	0
1985				
Low sulfur	.06	0	159.6	58.6
High sulfur	0	41.4	5.6	35.1
1990				
Low sulfur	0	0	177.5	54.6
High sulfur	17.2	76.8	36.5	88.0
2000				
Low sulfur	0	0	0	42.0
High sulfur	34.4	75.9	174	58.8

Note: Decline of consumption in 2000 is due to nuclear power expansion.

Table 4.29 Electricity prices under reference case and leasing moratorium (mills/kWh in current $)

	Demand region			
Case	1	2	3	4*J*
1990				
Reference case[a]	85.8	87.8	66.8	65.2
Leasing moratorium	85.9	87.9	67.5	70.2
2000				
Reference case[a]	125.1	143.8	117.7	103.0
Leasing moratorium	125.2	144.5	122.7	107.8

[a]The reference case is the BACT case.
[b]There is a shortage in this region. We have recalculated prices, assuming this shortage could be made up by producing additional electricity with oil.

5	6	7	8J	9
104.0	81.4	135.5	56.4	12.3
0	0	0	0	0
128.7	50.3	141.6	63.3	14.8
6.0	33.3	0	48.2	9.4
0	0	146.5	16.9	14.2
202.9	105.5	15.6	148	41.1
0	0	91.4	47.7	11.6
192	125.7	157.0	147.4	47.5

5	6	7	8J	9
77.6	60.7	65.9	62.2	69.9
77.4	60.7	67.6	64.2	70.5
125.0	91.5	121.7	104.5	119.9
125.1[b]	91.7	125.4	107.8	121.7

Table 4.30 Industrial coal use under a leasing moratorium (million tons of 22 million Btu/ton)

Year	Leasing moratorium	No-leasing moratorium
1980	96.9	97.2
1985	97.5	111.2
1990	107.5	131.7
2000	128.7	175.7

Note: This excludes metallurgical and export.

BACT is already an actual policy of the United States. It has the effect of increasing reliance on high-sulfur coal throughout the United States. Since the east consumes high-sulfur eastern coal, even without the leasing moratorium, the main change in consumption patterns comes in the midwest. The midwestern regions consume more eastern and less western coal as the price of their low-sulfur coal rises. Expansion of high-sulfur coal in the east in the long run can occur with little increase in cost. Therefore the leasing moratorium results in a moderate increase in the price of electricity in the midwest.

The biggest effect on the price of electricity comes in the west, because of the increase in high-sulfur coal prices. This is a surprising and an important result. BACT reduces the need for western low-sulfur coal. But driving up the price of high-sulfur coal through a moratorium has significant impacts.

The second insight we have gained is the tension that the leasing moratorium creates for other segments of the industry. We observed the need for a large expansion in eastern coal, particularly midwestern coal. The point to be made is that the system might not be able to respond to such pressure. The same environmental reactions could arise in the midwest as are now being experienced in the west. Additional nuclear capacity is also needed toward the end of the period, and at present the ability to build such capacity is problematical. In short we can substantially reduce our reliance on western coal only if we are willing to trade off for pressure elsewhere in the system.

The size of the necessary reallocation of production capability raises yet another issue. If leasing policy and other policies were clear, it is not unrealistic to expect that large expansion could take place. However, if policies are not clear, decisions will not be made, and expansion to the necessary levels will not take place. In that case costs will rise far above the levels estimated here. However, recent moves to clarify leasing policy are encouraging.

In sum there are several avenues of substitution that provide flexibility to the system and limit costs. If we cannot take advantage of these options because policy precludes them, costs will be higher. If lack of clear policy also retards adjustments, costs will also be higher.

The Grab for Rents

The central role of the west in coal development arises from the relatively low cost of production. Even allowing for the high cost of transport, western coal is the cheapest alternative west of the Mississippi River and is competitive with eastern coal in the midwest. Thus far we have assumed that a competitive structure characterizes this market. While this is a plausible assumption for coal producers, it is not tenable for other segments of the industry. Specifically the railroads and state legislatures are in a position to take advantage of the western coal's strong market position. Since western coal is the low-cost source over a wide area, they have a fair amount of leeway in raising coal prices.

The railroads and the state legislatures are competing for the same pie. The state legislatures have already increased taxes on coal production. Montana has imposed a 30 percent levy. Wyoming has imposed a 16.5 percent severance tax, and other states are discussing raising their tax rates. The railroads have also jumped on the bandwagon. The real rate of increase in unit train rates has been large, amounting to over 4 percent annually since 1970, and recent increases have far exceeded even that level.[27]

The coal model was used to calculate the tax on Powder River Basin coal that would maximize revenue for the states of Montana and Wyoming. That optimal tax was 75 percent of the base price, or $4.05 per ton of coal. The results are shown in figure 4.2.[28] That tax level seems quite high and far from current levels. However, 75 percent of mine-mouth price is equivalent to about a 30 percent increase in the 1977 rail rates to Chicago.

Tables 4.31 and 4.32 show the results of this increase in transport rates on regional production and allocation. The main effect is to shift production out of Montana and Wyoming into the midwest. The increased rents make up for this loss in volume. For coal originating in the Powder River Basin, total railroad rents increase from $750 million to almost $950 million in 1990, or 26 percent. By 2000 the percentage increase in rents is 43 percent, reaching $1,000 million per year, in current dollars.[29]

Regional allocation patterns are very sensitive to transport rates. It is to the advantage of railroads to increase rates as they have done. This in turn will discourage western output.

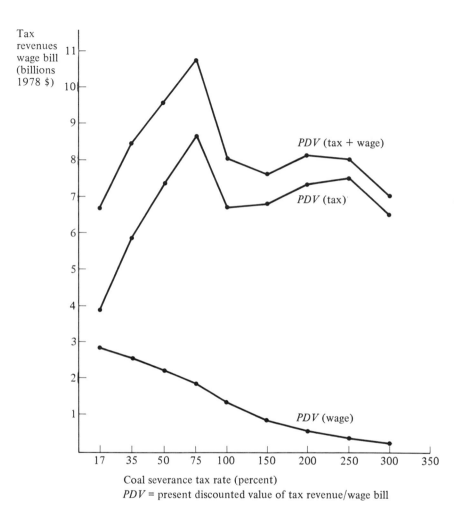

Coal severance tax rate (percent)
PDV = present discounted value of tax revenue/wage bill

Figure 4.2 Present value of Montana-Wyoming tax revenues as a function of tax rate for 1975 to 2000, with a 17 percent nominal discount rate. *PDV* = present discounted value of tax revenue wage bill.

Table 4.31 Regional output under western transport rate increase (30 percent above 1977 levels, in million tons)

Year	Region					
	1	2	3	4	5	6
1980	181.8	266.4	156.8	142.6	37.7	53.6
1985	222.4	257.8	83.8	289.8	65.1	85.8
1990	340.7	184.0	272.9	390.4	50.0	126.8
2000	208.0	182.7	495.9	480.6	67.7	84.7

Who pays the rents? We can get an idea by examining the effect of higher transport taxes on delivered coal prices (see table 4.33). The western consuming regions pay, for they use the majority of western coal. In the east the price of coal hardly changes at all. The railroads were quick to respond to the perceived market opportunity. This leaves little room for the states to increase taxes unless they begin a process of economic warfare, or unless they find ways to promote interregional collusion.[30]

In sum the grab for rents is having an important effect on western and eastern coal production. In terms of cutting back Montana-Wyoming output, it almost achieves the same effect as an indefinite leasing moratorium. Of course the implications for income distribution are quite different.

4.6 Environmental Trade-Offs and Coal Policy

The major environmental issues facing the coal industry are sulfur pollution regulations and the expansion of strip mining. These two areas interact. The pressure of sulfur regulations causes an expansion in western output. We have estimated that instituting sulfur regulations costs, in monetary terms, $7 billion per year. In addition it entails a 50 percent expansion in western (Montana-Wyoming) strip mining.

Best available control technology was mandated in an attempt to reverse this expansion, without loosening sulfur regulations. Our estimate of the costs of this policy are $3 billion per year. The effect on reversing western output growth is rather small and comes late in the period. Other factors, such as the rise in rail rates, will have even a larger impact on slowing western growth. Nevertheless, as we saw in the examination of basic forces, substantial western expansion will occur regardless of policy developments or factor price developments. The base level of expansion is primarily determined by demand growth west of the Mississippi River.

Table 4.32 Distribution of coal with high western transport rates in 2000 (million tons)

	Demand Region					
Supply region	1	2	3	4	5	6
1	36.36	81.68	0	0	5.97	0
2	0	0	0	0	21.74	3.96
3	0	0	174.72	0	175.74	122.35
4	0	0	38.99	110.87	0	0
5	0	0	0	0	1.28	2.97
6	0	0	0	0	0	0
Total	36.36	81.68	213.71	110.87	204.74	129.28

Note: Excludes metallurgical and export.

Table 4.33 Regional delivered coal prices under higher transport costs (current $/million Btu)

	Demand region			
Case	1	2	3	4*J*
1980				
High sulfur	1.44	1.33	1.32	1.17
Low sulfur	1.71	1.59	1.50	1.22
1985				
High sulfur	2.11	1.94	1.93	1.76
Low sulfur	2.82	2.64	2.46	1.78
1990				
High sulfur	2.95	2.72	2.66	2.33
Low sulfur	3.96	3.73	3.34	2.39
1995				
High sulfur	5.67	5.25	4.70	3.99
Low sulfur	7.49	7.10	5.67	4.04

7	8	9	10	11	12	Total
0	0	0	0	0	0	124.01
0	0	0	0	0	0	25.70
13.10	0	0	0	0	0	485.91
0	124.19	56.44	29.66	120.48	0	480.64
14.67	0	30.02	0	0	5.78	54.83
11.52	0	0	0	0	73.15	84.66
39.29	124.19	86.46	29.66	120.48	78.93	1,255.66

5	6	7	8J	9
1.41	1.26	1.43	0.78	1.38
1.64	1.43	1.46	0.79	1.41
2.16	1.90	2.07	1.08	2.01
2.75	2.49	2.19	1.15	2.08
3.02	2.61	2.84	1.46	2.76
3.88	3.55	3.14	1.64	2.85
5.77	4.62	5.46	2.77	5.22
7.27	6.70	5.85	3.49	5.26

The pressure against environmental regulations is now increasing, with the expectation that looser standards would expand coal use and lower oil imports. But we found that this method of reducing imports is quite costly. The logic is the following. At current prices oil use is disappearing in the electricity utility sector. Environmental policies that raise the cost of coal simply diminish the growth of electricity demand and increase the attractiveness of nuclear power.

The area where coal does compete with oil is in the industrial sector. Here increased coal prices due to environmental policy do increase relative oil use. However, the effect is small. Again, an important effect of higher coal prices is a reduction in energy demand. In sum only an extremely high value attached to import reduction would call for loosening environmental standards because of our oil import reduction goal.

The trade-offs involved are thus between sulfur pollution, strip mining, and the monetary costs of such policies. So far it appears that the nation has been willing to pay the price to clean the environment.

In all of our analyses nuclear power has expanded at the end of the century. After 1983 the model allows for the lowest-cost choice for future power generation. This assumption has been made purposely to obtain lower-bound estimates of the costs of coal policy. We are now ready to drop that assumption and examine the sensitivity of our estimates to the role of nuclear power. What does the coal industry look like, and what are the costs of prohibiting a major expansion of nuclear power?

References

1. Alt, C., and M. B. Zimmerman. "The Western Coal Tax Cartel." MIT Energy Laboratory working paper (forthcoming).

2. Baughman, M. L., P. L. Joskow, and D. P. Kamat. *Electric Power in the United States: Models and Policy Analysis.* Cambridge, Mass.: The MIT Press, 1979.

3. Data Resources, Inc. "New Source Performance Standards: The EPA Considers Its Options" (by S. E. Martin). *Coal Review* (December 1979): 17–46.

4. Dyck, D., et al. "Comparison of Fuel and Technology Alternatives for Industrial Steam Generation Systems." MIT Energy Laboratory working paper no. MIT-EL 79-060WP, December 1979.

5. Environmental Protection Agency. *Environment News* (July 1979).

6. Ramsay, W. *Unpaid Costs of Electrical Energy.* Study prepared for the National Energy Strategies Project. Baltimore, Md.: The Johns Hopkins University Press, 1979.

7. Tyner, W., and R. J. Kalter. *Western Coal: Promise or Problem.* Lexington, Mass.: Lexington Books, 1978.

8. U.S., Department of Commerce, Bureau of the Census. *Statistical Abstract of the United States.* Washington, D.C.: Government Printing Office, 1978.

9. U.S., Department of Energy, Energy Information Administration. *Analysis of Proposed U.S. Department of Energy Regulations Implementing the Power Plant and Industrial Fuel Use Act.* DOE/EIA-0102/21. Washington, D.C.: Government Printing Office, November 1978.

5   The Coal Industry without Nuclear Energy

Since there is a great deal of long-run substitution in the system, between low- and high-sulfur coal, coal and nuclear power, and producing regions within the coal industry, the environmental goals analyzed in chapter 4 are achievable at tolerable costs. However, what happens to these costs when policy actions reduce the possibilities for substitution?

In each of the previous scenarios nuclear power assumed a large role by the end of the century. In light of recent developments these levels of nuclear power will not be achieved. In fact a serious question arises as to whether new plants not already under construction or ordered will *ever* be built. Active public concern has led to legislation in several states prohibiting nuclear power plant construction until plans are developed for long-term disposal of spent fuel. This legislation has also been proposed at the federal level. Public acceptance of nuclear power in the aftermath of Three Mile Island is rapidly declining. All of these factors combined suggest that the substitution between nuclear power and coal is not likely to be realized. What happens to the cost of electricity, and what are the effects of a nuclear moratorium on the coal industry?

5.1   The Electric Utility Model for Predicting Nuclear Power Generation

Chapter 2 discussed the method used in the Baughman-Joskow-Kamat model to forecast nuclear power use. In essence the model compares the cost of base-load nuclear power to the cost of base-load fossil power in each region and then chooses the cheapest alternative. The same cost of capital is used to evaluate costs in both types of plant. The result of the choice is that the model, when unconstrained, builds a great deal of new nuclear power.

This result is consistent with other engineering studies of the cost of power, almost all of which conclude that nuclear power is a low-cost technology everywhere in the United States except in the mountain states. A recent study was performed in 1978 by the Nuclear Regulatory Administration of the Department of Energy; see Nuclear Regulatory Commission [4]. Table 5.1 reproduces its main conclusions. At a modest increase in coal prices in real terms, nuclear power proves to be the low-cost technology everywhere in the United States.

The actual choice of base-load plants in the United States is not consistent with the model and the data sketched here. Much less nuclear capacity has been built than either the Baughman-Joskow-Kamat model or the Federal Energy Regulatory Administration had predicted. Clearly other factors are at work.

Several factors could explain the divergence between the electricity model's predictions and actual investment behavior. First, for policy analysis, our assumptions about capital costs and capacity factors were overly optimistic. We altered these, raising costs of nuclear plants by 10 percent and lowering capacity factors to a 65 percent maximum achievable level.[1] Table 5.2 presents the results. Note that substantial amounts of nuclear power would still be built. This is not surprising since the Nuclear Regulatory Commission study used the same capacity factors

Table 5.1 Comparison of the cost of generating base load electricity by coal and nuclear power plants (present discounted value of costs for thirty-year operation in million $)

| Region | Nuclear | | Coal | |
	Nuclear with Pu recycle	Nuclear without Pu recycle	Constant real fuel cost	Escalating real fuel cost at 2 percent per year
New England	9,333	10,597	11,208	14,884
New York-New Jersey	9,508	10,772	11,032	14,510
Mid-Atlantic	8,717	9,981	9,782	13,051
South Atlantic	8,767	10,031	10,421	13,912
Midwest	9,943	10,558	10,164	13,368
Southwest	8,650	9,914	10,396	12,684
Central	9,203	10,467	9,684	12,732
North central	8,976	10,240	8,788	11,198
West	9,344	10,608	10,434	12,238
Northwest	9,276	10,540	10,843	12,939

Source: U.S., Nuclear Regulatory Commission, *Coal and Nuclear: A Comparison of the Cost of Generating Base-Load Electricity* (Washington, D.C.: Government Printing Office, December 1978).
Note: Assumes 5 percent inflation, 10 percent discount rate, 11 percent fixed charge rate, and 65 percent capacity factor.

Table 5.2 Sensitivity of generation capacity to reduced capacity factors and higher capital costs

Case	1980	1985	1990	2000
Nuclear generation capacity (GW)				
Reference case	64.5	122.8	125.4	387.9
High nuclear cost case	64.5	122.8	124.9	380.2
Coal generation capacity (GW)				
Reference case				
Coal with flue gas desulfurization	0	74.2	345.7	525.2
Coal without flue gas desulfurization	299.6	269.6	172.0	133.8
High nuclear cost case				
Coal with flue gas desulfurization	0	69.4	327.9	514.1
Coal without flue gas desulfurization	300.2	269.9	172.1	134.0

Note: High nuclear case assumptions have 10 percent higher capital cost and maximum achievable capacity utilization of 65 percent.

as this scenario. Note also that coal capacity declines. Again, higher costs of electricity lead to lower electricity demand and a decline in all types of base-load capacity.

There are other reasons why electric utility models predict such large amounts of nuclear power. These are essentially linear programming models that show large responses to small cost differences. All the engineering studies show a range of costs for coal and nuclear plants within a given region. Linear programming models, however, act as if there were a single cost for each, and therefore, when a range of outcomes is likely in any region, they obtain extreme results.

Furthermore electric utility models typically characterize decision makers as using the same cost of capital for both coal and nuclear plants. In reality this may not be true. The longer lags associated with nuclear construction mean that a utility must forecast demand over a longer-time horizon than for coal plants. The coal plant thus acquires additional value insofar as a commitment can be made later, when the utility knows more about demand in the year that the plant is expected to come on line.[2]

Finally, and most important, regulatory problems associated with nuclear power imply costs that are very difficult to quantify. Utilities might decide to pay the apparently higher costs associated with coal to avoid the hard-to-estimate, but nevertheless real, costs of regulatory delays and uncertainties involved with nuclear power construction.

Rather than attempt to model this complicated process, we adopt a simple expedient. We simply fix nuclear power construction. Earlier chapters assumed a partial moratorium on new announcements through 1983, which effectively constrained construction of nuclear plants through 1993. Here we make an assumption that the moratorium is kept in effect indefinitely. This policy is currently being given serious consideration and could in fact become actual policy. Further under the current environment no utility would consider building a nuclear plant.[3] Then we ask what the effects are on the coal industry and on the electric utility industry. We obtain a measure of cost consistent with a complete nuclear construction moratorium. In this manner we are able to assess the cost of the regulatory difficulties and uncertainties facing nuclear power.

It could be argued that these regulatory costs simply reflect the true social costs associated with waste disposal and safety issues. However, this is a complex and partially subjective question beyond the purview of this analysis. Instead we examine the effects of a total nuclear moratorium on the coal industry and on the price on electricity.

5.2 The Nuclear Moratorium

Under the moratorium on new nuclear construction, the effect on the production of coal is enormous. The size of the U.S. coal industry increases by over 50 percent by the end of the century. The regional impacts are also severe. Growth still occurs in the midwest and the Powder River Basin, with both regions producing at levels exceeding the current national output (see table 5.3).

Coal flows retain a pattern similar to the reference case; the same supply regions still supply each demand region. The reference case is the

Table 5.3 Coal production by region under a nuclear moratorium (million tons)

Year	Supply region					
	1	2	3	4	5	6
1980	201.1	251.5	129.8	192.7	13.5	62.1
1985	183.0	188.7	84.4	357.4	25.8	87.5
1990	322.0	196.2	177.1	557.9	63.5	105.0
1995	302.4	208.4	414.5	765.0	96.4	94.3
2000	294.9	188.0	741.4	1047.6	79.5	84.6

BACT case of chapter 4. The change is simply in the quantity of coal shipped (see table 5.4). A non-nuclear future implies a reliance on coal not currently appreciated.

The nuclear moratorium case begins to diverge from the reference case only after 1993. This is because our reference case includes a partial moratorium on new announcements lasting through 1983. In light of the developments following the accident at the Three Mile Island station, this is not at all unrealistic. Our base case already allows for slow nuclear growth through 1993. In essence public policy is now deciding the shape of the coal industry and the electric utility industry in the years 1995 to 2000. What this tells us is that a total moratorium on new nuclear construction will have an enormous effect at the end of the century.

The Effect of a Moratorium on Coal Prices
Under a total nuclear moratorium three factors will serve to raise costs in the coal industry. The increase in demand for coal will increase cumulative output and therefore costs. Second, as some of the remaining elasticity in the demand for coal is removed, railroads and state legislatures might see their market power expand. These groups could then raise the costs of rail rates and taxes. Lastly, as the rate of production expands, bottlenecks and short-run cost increases will begin to have an impact.

The short-run cost increase cannot be treated in our model, however. The model is a long-run equilibrium model and thus establishes a lower limit on cost increase. The conclusions about rents, discussed in chapter 4, still hold. The total appropriable rents will increase. However, the major factor limiting factor price increases was not nuclear power but interregional competition. The coal flow data of table 5.4 suggest the same pattern of regional competition will still be at work.[4]

Depletion due to cumulative output can be measured. Table 5.5 presents cumulative output by region with and without the moratorium, and table 5.6 presents a comparison of mine-mouth prices according to the sulfur intervals of chapter 2. The results emphasize the elasticity of the key segments of the U.S. coal supply curve. Costs do not increase a great deal because the Powder River Basin and the midwest have elastic cumulative cost curves. In other words, the impact of depletion elsewhere in the United States is limited because of these two key areas. A nuclear moratorium thus has its greatest impact on output in these two regions. The future of the electric utility sector is being built on these two regions.

The Effect of a Moratorium on Electricity Prices and Demand

The effects of a moratorium take time to be realized. Given the partial moratorium of the base case, the effects of the moratorium are not felt to any substantial degree until 1993.

Table 5.7 compares effects of a complete nuclear moratorium on electricity prices with a partial moratorium (rows *A* and *C*). The national average price of electricity in 1995 is 6 percent above the base case. Again, the price conceals a great deal of interregional variation. The hardest-hit regions are New England, the middle and south Atlantic regions, and the midwest. The mountain states are unaffected, since in all scenarios they choose coal for the base-load generation. This is not a surprise; the mountain states sit on vast low-cost deposits of coal.

The effect of the moratorium on the price of electricity is larger by the year 2000. The average increase is 12 percent, with a regional range of 0 to 26 percent. The latter increase is experienced in New England, which lies farthest from the coal fields and is most dependent on nuclear power.

By the end of the century the percentage increase in price is large, and the absolute cost staggering. The increase in electricity prices costs consumers $75.3 billion per year in 2000, or $24.5 billion in 1979 dollars. It is useful to relate this figure to other energy costs. The U.S. now imports approximately 8 million barrels of oil per day, or 2.9 billion barrels per year. At an approximate cost of $30 per barrel this amounts to a total cost of $87 billion. Thus the price of a nuclear moratorium is roughly equivalent to one-quarter of our total oil import bill.

Several biases in this cost calculation are working in opposite directions. On the one hand, because of favorable assumptions about nuclear power, we underestimate its cost. This results in an overestimation of the cost of a moratorium. On the other hand, we attribute no additional environmental cost to coal production other than basic reclamation costs. As discussed in the previous chapter, we could place significant costs on the social and aesthetic disruption entailed in this level of coal production. Also working to underestimate the cost of a moratorium is the long-run equilibrium nature of the model. Transition costs and divergences from long-run equilibrium are likely to be costly. Given the magnitude of the changes, these transition costs, in all probability, will dominate our estimates.

To obtain a lower bound on the cost and correct for our purposely biased assumptions, we compare the cost of a moratorium with the case of a lower nuclear capacity factor and 10 percent higher capital costs. In

Table 5.4 Coal distribution under a nuclear moratorium (million tons)

	1	2	3	4	5	6
Total shipments in 1990						
1	18.93	80.77	14.77	0.02	139.40	0.12
2	0.08	0.83	5.44	0.04	54.43	6.12
3	0	0	54.88	4.30	3.20	105.35
4	0	0	207.55	141.90	0	0.06
5	0	0	0.13	0.09	0	0
6	0	0	0	0	0	0
Total	19.01	81.61	282.76	146.36	197.03	111.65
Total shipments in 1995						
1	44.60	112.18	0	0	69.62	0
2	0	0	0	0	64.05	2.36
3	0	0	128.07	0	132.67	143.79
4	0	0	221.33	179.42	0	5.93
5	0	0	0	0	5.98	0
6	0	0	0	0	0	0
Total	44.60	112.18	349.40	179.42	272.32	152.08
Total shipments in 2000						
1	68.60	142.27	0	0	0	0
2	0	0	0	0	28.34	2.66
3	0	0	193.49	0	347.92	190.01
4	0	0	196.97	226.71	0	6.70
5	2.20	5.45	0	0	7.07	0
6	0	0	0	0	0	0
Total	70.80	147.72	390.46	226.71	383.33	199.37

Note: Excludes shipments of metallurgical and export.

Table 5.5 Cumulative coal output, 1975 to 2000, with and without a nuclear moratorium (million tons)

	Supply region					
	1	2	3	4	5	6
Moratorium	6,832.7	5,354.0	6,049.1	12,156.8	1,316.1	2,105.9
Reference case	6,317.9	5,253.3	4,472.7	10,966.5	1,275.8	2,061.8

7	8	9	10	11	12	Total
0	0	0	0	0	0	254.02
1.26	0	0	0	0	0	68.20
0	0	0	0.34	0	0	168.07
0	82.76	0	33.63	91.95	0	557.85
15.02	1.27	35.93	0	0	0	52.45
15.08	0.94	33.65	0	0	55.35	105.02
31.37	84.97	69.58	33.98	91.95	55.35	1,205.62
0	0	0	0	0	0	226.40
0	0	0	0	0	0	66.41
0	0	0	0	0	0	404.53
0	117.56	76.61	43.51	120.66	0	765.02
16.09	0	56.76	0	0	5.55	84.40
25.18	0	0	0	0	69.12	94.30
41.27	117.56	133.38	43.51	120.66	74.68	1,641.06
0	0	0	0	0	0	210.86
0	0	0	0	0	0	31.00
0	0	0	0	0	0	731.42
0	163.53	212.63	56.43	161.35	23.26	1,047.60
45.12	0	6.70	0	0	0	66.54
0	0	0	0	0	84.61	84.61
45.12	163.53	219.33	56.43	161.35	107.88	2,172.03

Table 5.6 Mine-mouth prices under a nuclear moratorium compared to reference case in 2000 ($/ton)

Sulfur category	Sulfur interval		
	1	2	3
Supply region			
Region 1			
Price under reference case	168.9	130.7	123.2
Price under moratorium	169.3	130.9	123.2
Region 2			
Price under reference case	197.1	129.6	122.6
Price under moratorium	197.1	129.6	122.6
Region 3			
Price under reference case	n.a.	164.0	105.1
Price under moratorium	n.a.	164.0	105.1
Region 4			
Price under reference case	29.7	27.9	27.8
Price under moratorium	29.7	28.6	28.4
Region 5			
Price under reference case	78.4	73.3	72.7
Price under moratorium	78.4	74.0	74.2
Region 6			
Price under reference case	102.7	72.4	71.4
Price under moratorium	101.7	74.7	74.1

the year 2000 the cost penalty due to a moratorium declines from a national average figure of 12 to 8 percent. The absolute cost thus declines to $18 billion (in 1979 dollars) per year in 2000, still quite a significant cost (see table 5.7).

These costs are large, but they may be worth the price. We do not know the true cost of making nuclear power acceptable to the public. The true social cost of waste disposal or of nuclear safety is at present unknown. There are those who would pay the price estimated here to avoid incurring those social costs. We do not attempt to assess this argument and can only point out that the estimates of this chapter indicate that the stakes are large.

The stakes are large in an additional way. The coal industry must expand enormously. The increase in environmental costs will be high. Further the build-up of carbon dioxide in the atmosphere at these levels of coal consumption could become an important factor.[5] Acid rain will be a more serious problem. In sum, if construction of new nuclear power plants is prohibited, we will pay a large cost in both monetary terms and potential environmental damage.

4	5	6	7	8
120.3	110.1	99.0	98.8	99.1
120.3	110.1	102.8	102.5	102.2
113.7	114.1	113.6	113.8	113.9
117.2	116.3	115.9	116.2	116.2
91.8	89.6	88.3	88.3	88.3
93.2	90.7	91.0	91.0	91.1
27.7	27.8	27.8	27.6	27.7
28.7	28.6	28.6	28.5	28.6
72.5	73.2	72.6	72.9	72.7
73.5	74.0	74.1	73.7	73.4
72.3	71.9	72.4	123.2	123.8
74.4	76.2	73.8	123.2	123.8

The Effect on Demand

The large increases in electricity price will have an important impact on the growth rate in the demand for electricity. In earlier chapters we saw that under a wide range of circumstances, growth in electricity demand averaged about 4.5 percent per year. This was far below actual experience before 1973, when annual demand growth averaged 7 percent. The impact of higher prices under a moratorium reduces annual electricity demand growth to 3.8 percent. Given the cost increase it entails, this implies a long-run elasticity of demand for electricity of roughly 1.0.[6] The important point is the enormous growth in the coal industry seen here comes even after adjusting for the 15 percent decline in absolute electricity demand caused by the nuclear moratorium.

5.3 Interaction of Coal and Nuclear Policy

Although the costs of rectifying environmental damage due to coal mining, under the assumption that nuclear power would supply significant amounts of electricity by the end of the century, would be significant, they could be borne by the large and flexible U.S. economy. But the costs

Table 5.7 Electricity prices under alternative scenarios (current mills/kWh)

	Demand region				
	1	2	3	4J	5
1980					
Case					
A	49.1	43.1	38.2	36.8	37.6
B	50.3	44.0	38.5	36.9	38.3
C	49.1	43.1	38.2	36.8	37.6
D	49.1	43.1	38.2	36.8	37.6
E	49.0	42.2	37.3	36.9	36.6
F	49.1	43.1	38.3	36.8	37.6
1990					
Case					
A	85.8	87.8	66.8	69.2	77.6
B	85.8	90.8	68.5	69.2	77.8
C	90.4	89.6	67.9	71.3	78.8
D	90.5	89.8	68.5	67.2	79.6
E	86.2	84.3	66.2	66.4	72.8
F	90.4	89.7	68.6	72.3	78.5
2000					
Case					
A	125.1	143.8	117.7	103.0	125.8
B	126.7	147.6	122.1	107.2	129.1
C	157.4	166.9	133.8	115.1	144.9
D	158.0	166.4	129.5	116.2	148.0
E	138.6	154.8	129.9	112.7	136.8
F	158.4	168.0	140.6	128.4	145.9

Note: A = reference; B = high nuclear costs; C = nuclear moratorium; D = nuclear moratorium with BACT; E = nuclear moratorium without sulfur regulations; F = nuclear.

6	7	8	9	National average
30.3	40.6	30.9	33.9	37.4
30.5	40.6	30.7	34.2	37.7
30.3	40.6	30.9	33.9	37.4
30.3	40.6	30.9	33.9	37.4
29.5	40.4	30.9	33.9	36.9
30.3	41.0	31.0	34.0	37.4
60.7	65.9	62.2	69.9	70.7
61.3	66.2	62.6	71.2	72.0
62.8	65.9	62.2	70.9	71.9
61.2	66.7	57.3	69.0	70.9
55.2	64.8	55.4	68.4	67.7
62.8	67.8	64.2	71.4	72.6
91.5	121.6	104.5	119.9	117.0
94.4	123.6	103.6	126.8	120.1
107.8	126.8	104.6	135.3	130.7
107.1	121.3	107.8	124.6	128.8
102.0	118.4	104.4	122.0	123.0
108.6	130.9	121.3	141.3	137.1

of doing without nuclear power will be large and will result in higher electricity prices to consumers. Under this condition there probably will be pressure to relax environmental standards. Trade-offs will undoubtedly be made between coal, the environment, and nuclear power.

If, as is now probable, a drastically reduced nuclear power industry becomes reality, the costs of clean coal will increase as output increases. Actions that increase the cost of coal cannot be mitigated by increasing nuclear capacity. We have already established a lower limit on the costs of a clean environment when we assumed an optimistic view of the ability of the nuclear industry to substitute for coal. Here we will calculate an upper limit. We allow no substitution.

The Effects of BACT in the Absence of Nuclear Power
What is the effect of BACT without the nuclear alternative? Or, without BACT, how much is the price of electricity lowered? Table 5.7, line D shows our estimates. Under a nuclear moratorium BACT raises the price of electricity by 1.5 percent in the year 2000. This is three times the increase experienced in 2000 without a nuclear moratorium. The impact in earlier years is the same, since the full moratorium differs from our reference case only toward the end of the century. In sum, the cost of lowering emissions through BACT rises when the nuclear option is no longer available.

The regional structure of effects remains the same under a full moratorium as under our reference case. The difference is that the effects are magnified. The increases in the price of electricity due to BACT are higher with a nuclear moratorium, but they still most heavily affect the western and midwestern regions, with little or no incremental cost in the east.

The magnitude of the changes in coal output is enormous, but the structure is the same with or without the moratorium. BACT is responsible for increasing midwestern output by 300 million tons by the year 2000 and reducing the output of the Powder River Basin by an equal amount. Table 5.8 shows the output results under a moratorium.

Again, we have modeled the extreme version of BACT. The most recent regulations lie in between. They favor midwestern coal to a lesser degree. They allow partial scrubbing of low-sulfur western coal. Since partial scrubbing is cheaper, some of the western markets will be preserved.[7]

The Cost of Sulfur Pollution Standards in the Absence of Nuclear Power
We have already seen that clean air regulations cost electricity consumers

Table 5.8 Regional coal production under a nuclear moratorium without BACT (million tons)

Year	Supply region					
	1	2	3	4	5	6
1980	201.1	251.5	129.8	192.7	13.4	62.1
1985	282.3	196.9	84.4	328.7	31.7	78.4
1990	323.5	190.6	141.0	576.7	92.9	78.4
1995	301.7	234.5	233.4	1007.1	76.1	59.9
2000	299.4	231.6	422.7	1301.0	127.0	80.7

over $6 billion per year (in 1979 dollars), if nuclear power is still a viable option. Table 5.7, line E compared to line A, shows the effects of sulfur standards when nuclear power is no longer an option. Removing sulfur standards lowers the national average price of electricity by 4.5 percent. This is in addition to the 1.5 percent reduction that would be realized by removing BACT. Thus the complete effect of imposing air pollution regulations is a 6 percent increase in the price of electricity. The absolute cost is $14.2 billion in (1979 dollars), over twice the cost of sulfur policy when nuclear power is a viable option. Again, the bulk of this cost comes from the sulfur regulations themselves and not from the BACT provision.

The regional pattern of electricity price increases is more evenly spread than when sulfur regulations are imposed in the absence of a nuclear moratorium. Prices increase everywhere. BACT raises western electricity prices. The east experiences price rises due to the higher cost of coal and, more important, due to the unavailability of nuclear power.

In sum, when nuclear power is not available, the costs of clean coal are high. Under these circumstances it is most probable that political pressure to reduce sulfur standards would intensify. The impetus would come not from a desire to increase the use of coal but rather from a political interest to keep prices down. There is a trade-off between two environmental goals: clean coals and no nuclear power.

Table 5.9 presents output by region when no sulfur standards are in effect. Comparing those numbers to the regional output shown in table 5.8, we see that the main beneficiaries of sulfur standards are again regions 4 and 5. The prime loser is again the midwest. However, by the end of the period the net difference is relatively small. Even with sulfur standards the midwest does relatively well under a full nuclear moratorium. And conversely, even without sulfur regulations the west must play an extremely important role.

Table 5.9 Regional output under a nuclear moratorium without sulfur standards and BACT

| | Supply region | | | | | |
Year	1	2	3	4	5	6
1980	323.0	193.0	117.5	130.2	13.5	62.3
1985	234.0	192.7	251.8	204.6	22.0	88.6
1990	344.2	194.7	348.7	336.7	69.0	102.6
1995	390.7	206.4	547.0	635.2	17.0	117.9
2000	338.6	221.6	549.4	1231.6	56.7	101.9

Western Coal in the Absence of Nuclear Power

A nuclear moratorium places tremendous pressure on western coal. Under all scenarios the west is called upon to expand production dramatically. What is the cost of restricting this development?

Table 5.7 provides a partial answer. By the year 2000, when a western leasing moratorium is combined with the full nuclear moratorium, there is a substantial increase in the price of electricity. With no nuclear moratorium national average electricity prices rose approximately 2 percent when leasing was halted; now they rise 5 percent. The total cost of electricity rises by $7 billion per year. The comparable figure without a nuclear moratorium is $3.0 billion per year (in 1979 dollars) in the year 2000. The regional breakdown of cost shows a pattern similar to the earlier leasing moratorium analysis. Prices rise most in the consuming areas that rely on western coal: the west and midwest. Western coal is important when nuclear power is available; it now becomes much more important.

The leasing moratorium combined with the nuclear moratorium does act to reallocate output. However, all it does is transfer the strain from Montana-Wyoming to the midwest. Again, the effects are felt quite late in the period. In fact, as table 5.10 indicates, the relative shifts are small through 1995.

The role of eastern Appalachian coal is interesting. Southern Appalachia remains priced out of the steam coal market under almost all circumstances. Northern Appalachia is able to expand considerably under the leasing moratorium, but it still represents a diminishing segment of the U.S. coal industry. The center of activity is shifting to the west and midwest. The shift is reinforced under the pressure of a nuclear moratorium. Environmental and transport policies with respect to coal will determine the split between the west and midwest.

The Effects on Demand

The policies analyzed here result in substantial increases in the price of electricity. Table 5.11 summarizes the effects of these policies on demand for electricity, and table 5.12 summarizes the effects on the derived demand for coal. Sulfur regulations, including BACT, are responsible for a 4.4 percent decline in electricity demand and a 2.4 percent decline in coal demand. A leasing moratorium alone would cause a 1.6 percent decline in electricity demand and a 3.0 percent decline in coal demand.

These results are not unexpected. Policies that increase the cost of electricity without affecting the price of coal have a twofold effect on coal demand. They decrease the demand for electricity and therefore the derived demand for coal. On the other hand, they increase coal demand in the industrial sector, where coal competes directly with electricity. This result is

Table 5.10 Regional output under a nuclear moratorium and a western leasing moratorium

	Supply region					
Year	1	2	3	4	5	6
1980	201.1	251.5	129.8	192.6	21.0	53.7
1985	201.7	276.7	85.9	344.7	50.8	57.6
1990	331.9	190.6	179.0	615.3	44.9	62.3
1995	311.3	211.1	455.6	801.5	70.5	10.3
2000	393.2	221.0	924.6	475.2	202.6	24.6

Table 5.11 The demand for electricity with alternative coal policies under a full nuclear moratorium (million MWh)

	1985	1990	1995	2000
Base case with nuclear moratorium	3127.6	3833.9	4570.5	5484.7
No BACT	3110.4	3832.6	4647.5	5575.3
No sulfur regulation (no BACT)	3148.5	3934.6	4775.1	5727.2
No leasing (with BACT)	3115.7	3814.8	4532.1	5394.4

Table 5.12 Coal consumption under various coal policies and a nuclear moratorium (million standard tons of 22 million Btu)

Case	1985	1990	1995	2000
Base case moratorium	904	1,263	1,715	2,221
No BACT	868	1,143	1,696	2,201
No sulfur regulation (no BACT)	900	1,276	1,773	2,275
No leasing (with BACT)	896	1,245	1,682	2,153

seen in the BACT case. The BACT policy results in a decline in electricity demand but an increase in coal demand. Leasing, however, affects coal prices directly and thus has a greater impact on coal demand than upon electricity demand.

There is another contributing factor to the lower relative decline in coal demand with BACT. Note that demand for coal in the years until 1990 actually increases with BACT. Since BACT raises the cost of using coal, we would expect a decline in demand. What is happening is that the higher heat rates associated with scrubbers cause the increase in demand. Scrubbers require energy, and their introduction means more coal must be used per kWh. Since electricity demand is very inelastic in the short run, electrical demand is relatively unaffected.

5.4 The Trade-Offs

The United States is currently determining the future of the coal and electric utility industries. Because construction lags are long, decisions made today will not be felt until the 1990s. The single most important decision affecting the coal industry has nothing to do with environmental controls, nor has it anything to do directly with the coal industry. The central policy decision to be made, from the standpoint of future coal production, is the role of nuclear power.

This analysis has simulated two alternative paths. The first reflects current nuclear power difficulties, but assumes that attitudes change and the situation improves by 1983. The second alternative assumes that public acceptance of nuclear power is not restored and that no new construction takes place. That future begins to diverge from the more moderate assumptions in the early 1990s; by 2000 it represents a very different world. Coal production must expand almost 50 percent even with a reduced demand for electricity. Furthermore the expansion is concentrated in key segments of the U.S. supply curve, so the effects of this enormous expansion are most deeply felt in the midwest and in Montana-Wyoming. The size of the needed expansion indicates the problems likely to arise in the absence of nuclear power. There is a serious question whether such a rate of expansion can be achieved. Bottlenecks are bound to arise, particularly if decisions regarding nuclear power are postponed, making the future of coal uncertain. Reaction from people concerned with environmental degradation will certainly be great. In sum, prices of both coal and electricity could rise far above what is estimated here.

A system placed under the stress envisioned in these scenarios will face pressure from other groups within the economy. Besides pressure from environmental groups there will also be pressure from groups that think the costs of a clean environment are too high. We have seen that without nuclear power the cost of environmental policies rises. This is due to the larger role coal must play in the economy. Given the much higher costs of electricity, there will be groups that feel the higher costs of environmental policies are not worth the benefits. In sum, a coal industry without nuclear power necessitates some very difficult trade-offs. In our system of decentralized decision making, where diverse groups wield great influence, the path by which we arrive at those trade-offs is not likely to be smooth.

We have seen that all these policies have little impact upon oil imports. The very high cost of oil has eliminated it from the utility sector under most circumstances, and what we do in the utility sector has little impact on oil consumption. Of course, if the needed coal capacity cannot be built for reasons we discussed, oil will be used as a last resort.

We have driven the system to the brink. We have seen that current policy leads to extreme results. The absence of a nuclear alternative forces massive reliance on Powder River Basin and Illinois Basin coal. Only if we are willing to tolerate this expansion does a nuclear moratorium make sense. The perceived costs of nuclear safety and waste disposal must be traded off against the direct costs and environmental damage of a vastly expanded coal industry.

References

1. Data Resources, Inc. *Energy Review* (Winter 1979).

2. National Research Council. *Energy and Climate.* Washington, D.C.: National Academy of Sciences, 1977.

3. U.S., Department of Energy, Energy Information Administration. *Annual Report to Congress 1978,* vol. 3. Washington, D.C.: Government Printing Office, 1979.

4. U.S., Nuclear Regulatory Commission. *Coal and Nuclear: A Comparison of the Cost of Generating Base Load Electricity.* Washington, D.C.: Government Printing Office, December 1978.

5. Zimmerman, M. B., and R. P. Ellis. "What Happened to Nuclear Power?" MIT Energy Laboratory working paper 80-002WP, January 1980.

6. Zimmerman, M. B., and C. Alt., "The Western Coal Tax Cartel." MIT Energy Laboratory working paper (forthcoming).

6

The Future of Coal and the Alternatives

6.1 Economic Forces Influencing the Industry

The main supply and demand influences shaping the U.S. coal industry can be viewed as in a tug-of-war. A set of forces is causing movement toward the west. These forces dominate the current scene. Pulling in that direction are sulfur regulations (excluding BACT), labor cost trends, and shifting demand centers. Sulfur regulations cause an expansion in low-sulfur western output. Rising labor costs cause a shift at the margin to strip mining. Depletion of strip reserves in the eastern United States means that the shift to strip mining is also a shift to western mining, where strip reserves are plentiful and relatively low in cost. Both these trends are reinforced by the emergence of the states west of the Mississippi River as coal demanders. This change in demand patterns provides a base level of demand for western coal that implies, even in the absence of other factors, a healthy growth rate for western coal production. These trends occur even under extremely optimistic assumptions about nuclear power.

The developments pulling eastward are policy related. Regulations mandating best available control technology (BACT) for sulfur removal cause the midwest to reduce its reliance on western coal. The attempt to capture rents through higher taxes and railroad rates also slows the expansion of western coal. Table 6.1 summarizes these scenarios by presenting cumulative output from the west for each of the cases considered in chapters 3 and 4.

This tug-of-war will determine how much western coal moves east. Our analysis suggests that the key area in which this struggle takes place is the midwest. The east primarily uses eastern coal after BACT is imposed. The tug-of-war will determine how much western coal crosses the Mississippi River into the midwestern market.

Production patterns also change within the eastern producing regions. In the east, southern Appalachia exits from the steam coal market. Production of low-sulfur southern Appalachian coal for the metallurgical

Table 6.1 Cumulative output by region, 1975 to 2000 (million tons)

Case	Supply region					
	1	2	3	4	5	6
1	7,162	4,833	8,135	8,596	1,033	2,162
2	6,227	5,325	3,939	13,865	1,417	1,412
3	6,155	5,168	3,810	12,306	1,393	1,432
4	6,318	5,253	4,473	10,967	1,276	2,062
5	6,480	5,300	5,068	10,492	1,363	1,024
6	6,652	5,344	5,838	7,493	1,667	2,084
7	6,832	5,354	6,049	12,157	1,316	2,106
8	7,032	5,415	6,827	11,187	1,689	1,054

Note: Key to cases: (1) no factor price changes, no sulfur regulations; (2) no factor price regulations, 1.2 lb SO_2 standard; (3) base case factor price change, 1.2 lb SO_2 standard; (4) base case factor price changes, 1.2 lb SO_2 standard, BACT required; (5) same as 4, with western coal land-leasing moratorium; (6) same as 4, with 25 percent real increase in western transport rates; (7) same as 4, with nuclear moratorium; (8) same as 7, with western land-leasing moratorium.

and export markets drives up costs enough to eliminate the region from steam markets. Both northern Appalachia and the midwestern producing regions compete for markets traditionally supplied by southern Appalachia. The competition between these regions is close, as shown in chapter 3. In sum low-sulfur eastern coal is reserved for premium markets, and high-sulfur eastern coal responds to growing demand in the electric utility sector.

The results of our analyses indicate that coal will be used primarily by electric utilities. At present the electric utility sector accounts for 75 percent of coal consumption. This fraction is likely to rise. In the case of a complete nuclear moratorium it would rise as high as 82 percent.[1] Thus the tremendous expansion in coal production occurs in order to supply an electric utility sector that is growing, by historical standards, at a low rate.

Industrial use of coal grows but still remains a relatively small fraction of total coal consumption. The historical decline of coal consumption by the industrial sector will be reversed, but it will remain a sector of lesser relative importance to the coal industry. The estimates here of industrial coal use are on the low side. We have not been able to reflect the impact of the Fuel Utilization Act of 1978, which prohibits construction of new gas or oil boilers. However, the model estimates a decline of oil and gas consumption, indicating that economics alone are sufficient to prevent new gas and oil boiler construction.

To obtain an idea of the range of possible levels of industrial consumption, we compare estimates from several of our scenarios to the Department of Energy forecasts. Table 6.2 makes this comparison. It is clear that, even if we take the upper range of industrial demand forecasts, the industrial sector will account for a fraction of output far below that of the utility sector.

The other areas of coal demand that will account for some of the coal expansion are domestic metallurgical coal, metallurgical and steam coal exports, and synthetic fuel needs. We have treated the first three of these sectors as exogenous, simply forecasting their needs and forcing the model to meet that level of demand. For metallurgical coal this is an adequate treatment. As discussed in chapter 2, world steel production is the primary determinant of demand for metallurgical coal; a model of steel production is outside the scope of our study.

The latter two demand categories will depend upon coal prices. However, a complete model of these sectors would have to consider international fuel markets, as well as domestic policy with regard to synthetic fuel production. All we attempt here is to examine the issues involved to obtain estimates of the role synthetic fuels and steam coal exports are likely to play.

Table 6.3 presents current cost estimates for synthetic fuels derived

Table 6.2 U.S. industrial coal consumption in selected scenarios compared to Department of Energy forecasts (million tons of 22 million Btu)

Case	1985	1990	2000
Reference (base case factor price increases 1.2 lb SO₂ standard, BACT)	100	114	155
Reference assumptions plus high western transport rates	95	107	140
Reference assumptions plus nuclear moratorium	100	145	167
Department of Energy	128	200	307[a]

Note: These totals exclude coal produced for domestic metallurgical use.
[a] Forecast for 1995.

Table 6.3 The estimated cost of synthetic fuels (1980 $/million Btu)

Fuel	Feedstock cost	Total cost
Coal liquids	1.20–1.70	5.40–9.25
Methane (from coal)	1.90–2.70	8.10–10.75
Shale oil	2.50	4.70–5.10

from coal. At present these fuels are not economic, although given the rapid escalation in real oil prices they will become economic in the future. The rate at which synthetic fuels will be produced in this country in the next two decades will depend upon government policy, particularly the level of subsidy. Congress has provided $20 billion in initial subsidies.[2] The Department of Energy sees a total coal demand from a synfuels industry of 28 million tons in 1990, 131 million tons in 1995, and rapid growth thereafter. They forecast coal demand for synthetic production of approximately 190 million tons by 2000. Even at the DOE projection, that is less than 9 percent of DOE-forecasted coal production in the year 2000.[3] In sum synthetic fuels are arriving, but their relative impact will not be large until the next century. In the next twenty years production of coal for synthetics is not likely to have a large impact on coal price.

A similar conclusion applies to steam coal exports. At present the United States exports relatively small quantities of steam coal. In the 1960s steam coal exports were significant but declined, as oil replaced coal in Europe and Japan. The high price of oil is reversing this trend. Nevertheless, the U.S. faces competition from other coal-producing nations including, among others, Canada, South Africa, Australia, Poland, and Colombia. How much coal is the United States likely to export for steam purposes? Table 6.4 presents some recent forecasts. Clearly by 2000 U.S. steam coal exports could be an important phenomenon, although how important is difficult to tell at this point. Coal for synfuels as well as steam coal exports will, because of cost, be western coal. Thus the incremental impact on coal prices will be small.[4]

6.2 Coal Policy and Oil Imports

Oil imports dominate all other issues in the energy policy debate. Our integrated coal and electric utility model allows a consideration of the trade-offs between environmental policy and oil imports.

Table 6.5 summarizes the results of various scenarios considered in this study. The imposition of sulfur standards has the largest impact, increasing oil consumption in the year 2000 by over 345 thousand barrels per day. Thus clean air requirements will lead to a moderate increase in oil imports.

The other policies considered here, however, have relatively small impacts on oil consumption: BACT, a leasing moratorium, transport rate increases, and a nuclear moratorium. This is an important result that should put the current policy debate into perspective. Although there are

Table 6.4 Forecasts of steam coal exports

Source	1985	1990	1995	2000
International Energy Agency (IEA)	13			59
U.S. Department of Energy (DOE)	20–35	22–42	24–49	
WOCOAL	20–35	30–60		65–130

Sources: IEA: OECD. *Steam Coal Prospects to 2000,* Paris 1978. DOE: Energy Information Agency. *Demand For World Coal Through 1995,* May 1979. WOCOAL: World Coal Study, Carroll Wilson, Dir. *Coal Bridge to the Future.* Cambridge, Mass.: Ballinger Press, 1980.

Table 6.5 Oil consumption under selected scenarios (10^{16} Btu/year)

Case	1985	1990	2000
A			
Residential, commercial, industrial	0.5584	0.5669	0.5863
utility	0.1235	0.2006	0.0658
B			
Residential, commercial, industrial	0.5637	0.5807	0.6036
utility	0.1246	0.2106	0.1242
C			
Residential, commercial, industrial	0.5627	0.5793	0.6051
utility	0.1232	0.1629	0.0987
D			
Residential, commercial, industrial	0.5654	0.5855	0.6146
utility	0.1231	0.1627	0.0912
E			
Residential, commercial, industrial	0.5640	0.5826	0.6186
utility	0.1232	0.1628	0.0925
F			
Residential, commercial, industrial	0.5627	0.5798	0.6302
utility	0.1232	0.1625	0.1215

Note: Cases: (A) base case factor prices, no sulfur regulations; (B) base case factor prices, 1.2 lb SO_2 regulations; (C) base case factor prices, 1.2 lb SO_2 standard, BACT; (D) same as C with 25 percent real increase in western transport rates; (E) same as C with moratorium on coal leasing; (F) same as C with a nuclear moratorium.

valid reasons why one might be against any of the policies discussed, *their effect on oil imports should not be one of those reasons.* Once having decided that sulfur regulations are worth the cost in both monetary and in oil import terms, the rest of the coal policy debate could ignore effects on oil imports.

The reason for this relatively small impact of coal policy on oil imports is as follows. As we have seen, the main consumer of coal is the utility sector. Even at the oil prices used in our simulations, oil is priced out of the utility market. The increased cost of coal due to environmental regulations does not close the gap between oil and coal prices. Oil disappears from utility boilers over the long run.

Higher coal prices affect oil imports primarily in the industrial sector. Two factors are at work. Higher coal prices cause substitution away from coal. But at the same time higher energy prices slow growth in the demand for energy as a whole in the industrial sector. Thus cutbacks in coal consumption are greater than the equivalent increase in oil consumption.

Consistent with this logic, oil price rising above the levels used in our analysis should have relatively small impacts on coal demand. To test this we ran the reference scenario with oil prices rising in 1980 to $25.47 per barrel and then continuing to grow at 3 percent per year in real terms. The result is to increase consumption of coal by 66 million tons of 22 million Btu. Thus a doubling of oil price yields an increase of 4.6 percent in coal demand, implying a cross-price elasticity of 0.046. This low cross-price elasticity explains why coal prices have generally not responded to the 1979 to 1980 increase in oil prices.

Given sulfur regulations then, the focus of debate turns away from oil imports to the other trade-offs that must be made. Our analysis has isolated three key elements in coal policy: western coal, eastern high-sulfur coal, and nuclear power. These are the three safety valves in the system. If we remove one element, we put pressure on the remaining two. So long as the other two elements are available, the costs of any policy initiative are moderated.

6.3 The Three Safety Valves and the Question of Timing

Each safety valve plays an important role. In chapter 4 we examined what happens when the availability of western coal is drastically reduced. Coal prices and electricity prices rise substantially in the west. Huge demands are placed on midwestern coal, scrubbing technology, and nuclear power. In chapter 5 we saw how much more exaggerated the impacts are when nuclear power is not available. The midwest is asked to produce at a level equal to 230 percent of current national output. The long-run equilibrium assumptions of this model become highly questionable at this level of expansion. The level of scale-up is so large that environmental obstacles will increase. Factor prices will rise. In sum the system comes under strong pressures that take a long time to dissipate.

These results say something important about long-term coal strategy. Environmentalists favor an eastern underground strategy. Such a strategy is costly when nuclear power is available. Without nuclear power, it becomes untenable. Prices are increased and potential environmental problems become very serious, as discussed in chapter 5. Trade-offs must be made between environmental goals. Even at moderate rates of growth in electricity demand we cannot avoid hard choices.

These results are not dependent upon any particular assumptions we have made. The conclusions are robust. A smaller growth rate in electricity demand would postpone the day of reckoning, but not by much. Recall that in the nuclear moratorium case, electricity demand growth is only 3.7 percent per year. More pessimistic assumptions about nuclear power would only narrow the difference between our no-moratorium and full-moratorium cases; they would not alter the conclusion about trade-offs. Without nuclear power the coal industry will have to expand enormously.

To test the robustness of our conclusion, we altered our GNP growth rate assumptions. The entire analysis has been conducted assuming a rate of growth in GNP of 3.8 percent. In the 1970s, generally perceived as a slow growing decade, real GNP managed to grow at an average annual rate of 3.2 percent. What if the next twenty years are, as some fear, a period of even slower growth? Will that reduce the acute nature of the trade-offs? To answer that question, we simulated a scenario equivalent to the reference case except that real GNP grows at an average annual rate of 2.5 percent to the year 2000. Table 6.6 shows the resulting generation mix. Generating capacity is down, but there is still a large amount of necessary nuclear and coal capacity. Coal consumption declines by 200 million tons. But required nuclear power has diminished only slightly. A nuclear moratorium would require massive expansion of coal outputs, even at a growth rate low by historical standards.

The estimate of a 200 million ton decline in coal output presents an upper-bound estimate of the decline. Low growth diminishes demand, but higher oil prices raise coal demand. Higher nuclear costs also raise coal demand. To examine the interaction of all these developments, we ran a case combining high oil prices, low growth, and the higher nuclear capital costs. The result, as shown in table 6.6, is a decline in coal demand of 3.06 quads, or approximately 140 million tons of coal. A nuclear moratorium added to this latter scenario still calls for a 50 percent increase in coal output (table 6.6). Recent developments in oil markets, the consequently lower macroeconomic growth forecasts, and higher nuclear power costs do not alter our basic conclusions.

Trade-offs are inevitable. The trade-offs to be made are between western low-sulfur coal, eastern high-sulfur coal, and nuclear power. The desire for independence from foreign oil should not confuse the issue. The issues of realistic concern are nuclear safety and waste disposal problems versus increased strip mining and the increased use of high-sulfur coal, with potential problems from atmospheric buildup of CO_2 as well as acid rain.

At present no centralized decision maker is weighing these alternatives and balancing one objective against the other. The current implicit moratorium on nuclear power plant construction results from the enormous public reaction against nuclear power and the consequent delays and regulatory hurdles involved in plant licensing and construction. The results of this moratorium will be enormous pressure on the coal industry.

All the problems we have noted here begin to manifest themselves in the 1990s. The lags in the system are long, and actions taken today are not likely to have effects until the 1990s. Consequently the problem is not

Table 6.6 Generating capacity and coal consumption under alternative assumptions in 2000

Case	FGD	Coal (GW)	Nuclear	Steam coal consumption $(10^{15} Btu)^a$
1	525.2	133.8	387.9	32.1
2	430.2	133.8	363.0	27.3
3	558.6	134.2	382.6	33.6
4	436.4	133.8	351.4	29.0
5	659.7	134.1	104.2	42.9

Note: Key to cases: (1) reference, see appendix B for assumptions; (2) low growth; same as 1 except GNP growth is 2.5 percent per year; (3) same as 1 except oil price rises in 1980 to $25.47 per barrel and increases by 3 percent real per year thereafter; (4) reference case except growth as in 2, oil prices as in 3, nuclear capital costs increased by 10 percent, capacity factor for nuclear plants is 0.65; (5) same as 4 with a nuclear moratorium.

[a] Steam coal consumption excludes metallurgical plus export coal.

likely to be perceived until even later. The danger is that by the time the problem is perceived, it will be too late to take corrective action.

The clearest example of this is the problem of nuclear power. It takes at least ten years to build a nuclear plant. Any nuclear plant that will be available in the 1980s has already been announced. A moratorium thus will affect electricity supply in 1990 and beyond. The public reaction against nuclear power means that few, if any, plants will be ordered in the next five years. In the interim some reactor vendors are bound to exit from the business, creating a serious question about the ability to resume nuclear construction five years hence, if attitudes change and it becomes apparent, as our results indicate, that nuclear power is needed.

We are determining today the shape of the coal industry and the electric utility industry in the years beyond 1995. Trade-offs must and will be made. The choice is whether we make policy after consciously weighing alternatives or whether we back into a future made bleak by a lack of foresight.

Reference

1. U.S., Department of Energy, Energy Information Agency. *Annual Report to Congress 1978,* vol. 3. Washington, D.C.: Government Printing Office, 1979.

Appendix A   Distribution of Strippable Reserves in Appalachia

Appalachian coal reserves occur in hills (see figure A.1).[1] The Bureau of Mines calculates strippable coal reserves as all coal lying under less than 120 feet of overburden.[2] The average contour of the area is used to calculate the distance before the 120-ft limit is reached. Then, given seam thickness and areal extent of the seam, it is a simple matter to calculate total tons.

The characterization of the Bureau of Mines, as shown in the figure, leads to a uniform distribution of reserves according to overburden ratios. There is one complication, however, in that, regardless of seam thickness, the distribution is truncated at the 120-ft limit. This means that maximum overburden ratios will be lower for thicker seams than for thin seams.

Consider all the coal at a given overburden ratio, $\bar{R}$. We will add all the coal in seams such that $h/t \leq \bar{R}$. If $t \leq 120/\bar{R}$, we will add all coal in that seam. If $t > 120/\bar{R}$, then some of the coal in the seam will exceed the

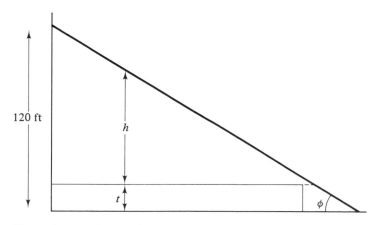

Figure A.1 Typical configuration of strippable reserves in Appalachia

120-ft overburden limit before $\bar{R}$ is reached. Thus the distribution of coal according to overburden ratio consists of two parts, both of which depend upon seam thickness.

Let

X_T = total strippable reserves,

ϕ = angle of contour,

t = seam thickness in feet,

$\bar{R}$ = given overburden ratio in feet, h/t,

$X_{t,\phi\bar{R}}$ = total strippable reserves in seams of t, with angle ϕ, available at less than $\bar{R}$,

Z = tons per acre foot,

F = acres of deposit,

Then

$$X_{t,\phi,\bar{R}} = \frac{t\bar{R}}{\tan \phi} \, tZF, \quad \text{if } t\bar{R} \le 120$$

$$= \frac{120}{\tan \phi} \, tZF, \quad \text{if } t\bar{R} > 120.$$

Thus

$$\iint_{t \ \phi} X_{\phi,t,\bar{R}} \, dt d\phi = X_{\bar{R}} \int_{120/\bar{R}}^{\alpha} \int_{0}^{90} \frac{120}{\tan \phi} \, tZFd\phi dt + \int_{0}^{120/R} \int_{0}^{90} \frac{t\bar{R}}{\tan \phi} \, tZFd\phi dt$$

$$= \int_{120/\bar{R}}^{\alpha} X_t \, dt + \int_{0}^{120/\bar{R}} \frac{t\bar{R}}{120} \, X_t \, dt. \tag{A.1}$$

Now

$$X_t = \left[\int_{0}^{90} P(t|\phi)p(\phi) \, d\phi \right] X_T$$

$$= P(t)X_T. \tag{A.2}$$

where

$P(t)$ = probability of observing a ton of coal coming from a seam of thickness t.

Substituting (A.2) into (A.1) yields

$$X_{\bar{R}} = X_T \left(\int_{120/\bar{R}}^{\alpha} P(t) \, dt + \frac{\bar{R}}{120} \int_{0}^{120/\bar{R}} tp(t)dt \right).$$

We take $P(t)$ as lognormal with μ and σ as estimated in chapter 2. Numerical methods are used to calculate the distribution, $\bar{R}$.

Reference

1. Zimmerman, M. B. "An Economic Interpretation of Coal Reserve Estimates." In *Resource and Reserve Estimation,* edited by M. A. Adelman et al. Cambridge, Mass.: Ballinger Press (forthcoming)

Appendix B   Base-Case Assumptions in the Coal Model

Table B.1 Tax rates

State	1976	1978 and beyond
Maryland	0	0.6
Ohio	0.2	a
Pennsylvania	0	a
Northern West Virginia	3.85	a
Alabama	0.52	1.3
East Kentucky	4.5	4.25
Tennessee	0.8	a
Virginia	0	1.0
Southern West Virginia	3.85	a
Illinois	0	5.0
Indiana	0	a
West Kentucky	4.5	4.25
Montana	30.0	35.0
Wyoming	16.5	16.8
Colorado	3.0	a
Utah	0	a
Arizona	0	2.5
New Mexico	3.54	3.8

Note: Current severance taxes, which are actually flat charges per ton, were converted to percentage increases to maintain constant relative relationships over time.
[a] No change.

Table B.2 Assumed real rates of escalation

Variable	1977–1980	1981–1985	1986–2000
Wages	0.10	0.02	Constant
Capital costs	0.02	0.02	Constant
Rail rates	0.03	0.03	Constant

Table B.3 SO$_2$ constraints on incremental coal (lb/million Btu)

Demand region	1975	1980	1985
1	3.0	2.0	1.2
2	3.15	2.05	1.2
3	4.16	2.39	1.2
4	4.58	2.52	1.2
5	2.83	1.93	1.2
6	4.36	2.46	1.2
7	1.20	1.20	1.2
8	1.20	1.20	1.2
9	1.38	1.30	1.2
10	2.67	1.87	1.2
11	1.35	1.28	1.2
12	1.27	1.24	1.2

Table B.4 Btu per ton of coal (10^6 Btu)

Region	Level
1	23.6
2	23.3
3	21.7
4	17.0
5	21.7
6	19.8

Table B.5 Exogenous levels of metallurgical and export coal (million tons)

Year	Metallurgical	Export	Total
1980	110	66	176
1985	116	77	193
1990	124	90	214
1995	132	105	237
2000	140	122	262

Table B.6 Reference case assumptions of the electric utility model: expected capital costs for plants (current $/kW)[a]

End of year[b]	Type				
	FGD	Gas	Oil	Coal	LWRU
1975	493.6	390.1	430.4	475.6	534.1
1985	1,187.0	874.5	1,030.5	1,047.4	1,278.6
1995	2,035.8	1,443.9	1,769.1	1,796.6	2,212.1

[a]Includes expected real cost increases during construction. The expected cost comes from applying the moving average algorithm to realized costs, as explained in the text. Realized costs are used to calculate the rate base. (In 1975, for example, new nuclear units are expected to cost $534 for kilowatt. For units coming on stream in 1975, the actual cost is a weighted average of realized costs over the previous ten years.)

[b]For convenience of calculation the rate of inflation is assumed to be 5.5 percent per year from 1976 to 2000. Model relationships are all in real terms, and these real values can be recovered by deflation of 5.5 percent.

Table B.7 Reference case assumptions of the electric utility model: realized capital cost ($/kW)

	1975	1985	2000
FGD	432.92	886.24	2,124.36
Gas	280.40	516.40	1,153.40
Oil	340.20	714.00	1,634.40
Coal	378.38	745.54	1,810.26
LWRU	471.40	996.80	2,316.60
IC-GT	134.00	250.80	589.94
Hydro	185.96	478.92	1,305.20

Table B.8 Reference case assumptions of the electric utility model: regional capital cost multipliers

Region	Fossil	Nuclear
1	1.0	1.0
2	1.0126	1.0
3	0.986	1.0
4J	0.9531	0.9359
5	0.9527	0.9515
6	0.8760	0.8510
7	0.8718	0.8449
8J	0.9097	0.9081
9	0.9601	1.100

Table B.9 Reference case assumptions of the electric utility model: heat rates (Btu/MWh)

	1975	1985	2000
FGD	10.89	11.10	11.08
Gas	10.08	10.58	10.72
Oil	10.08	10.58	10.72
Coal	10.08	10.28	10.26
LWRU	10.60	10.37	10.25
IC-GT	16.00	13.71	12.80
Hydro	10.08	10.03	9.95

Table B.10 Reference case assumptions of the electric utility model: operation and maintenance costs (current $, mills/kWh)

	1975	1985	2000
FGD	2.38	4.50	10.20
Gas	1.84	3.48	7.87
Oil	2.11	4.02	9.06
Coal	1.66	3.12	7.13
LWRU	1.24	2.27	5.19
IC-GT	2.85	5.23	11.11
Hydro	0.70	2.28	5.62

Table B.11 Reference case assumptions of the electric utility model: maximum capacity factor achievable

	Duty cycle	Availability factor
FGD	0.81	0.95
Gas	0.96	0.95
Oil	0.96	0.95
Coal	0.96	0.95
LWRU	0.86	0.85
IC-GT	0.15	0.99

Note: The product of the columns gives the maximum capacity factor achievable. The model solves for the most efficient utilization pattern.

Table B.12 Reference case assumptions of the electric utility model: other fuel prices

	Oil (current $/bbl)	Gas (current ¢/mcf)	U_3O_8 (1975 $/lb)	Unit separative work cost (constant 1975 $/swu)
1975	9.41	164.0	12.51	75.70
1980	17.18	212.0	—	—
1985	26.53	417.0	32.368	85.0
1990	39.06	819.95	—	—
1995	58.05	124.75	—	—
2000	85.70	182.62	53.022	85.0

Note: Oil prices reflect developments through June 1979. Since we have multiplied real dollars by 5.5 percent, our numbers for oil and gas are too low compared to 1979 dollars. Inflation between 1975 and 1979, as measured by the GNP deflator, was 6.8 percent per year on average. Thus, to convert to actual dollars before 1980, multiply our numbers by $(1.013)^t$.

Table B.13 Reference case assumptions of the electric utility model: fuel transport charges

Region	1975	1985	2000
Oil (current $/mmBtu)			
1	− 3.0	− 7.70	− 17.30
2	0	0	0
3	17.00	23.20	51.86
4J	5.44	15.50	34 58
5	− 6.00	− 15.50	− 34.60
6	1.00	− 7.70	− 17.30
7	1.00	− 7.70	− 17.30
8J	12.00	30.96	69.15
9	18.00	46.40	103.64
Gas (current ¢/mcf)			
1	37.2	77.36	172.78
2	28.8	54.16	120.88
3	28.8	54.16	120.88
4J	− 11.48	− 38.70	− 86.34
5	0.60	− 15.50	− 34.60
6	18.24	38.70	86.33
7	− 1.20	− 7.70	− 17.30
8J	− 1.20	− 7.70	− 17.30
9	27.20	54.16	120.88

Table B.14 Reference case assumptions of the electric utility model: financial parameters

Parameter	Percent
Regulated return on equity	0.16
Cost of debt	0.085
Cost of preferred stock	0.087
Debt limit	0.60
Minimum interest coverage ratio	2.0
Preferred stock function	0.10

Note: The rate of growth in GNP and value added in industry is 3.8 percent per year in real terms. Inflation is assumed to be 5.5 percent per year.

Table B.15 Reference case assumptions of the electric utility model: nuclear capacity constraints (maximum GW committed/year)

	1977	1980	1985	1990	1995	2000
Region 7	0.367	0.905	1.801	8.70	19.60	30.2
All others	2.5	5.350	10.10	15.60	21.60	30.2

Appendix C   Strip-Mine Reclamation Costs in the Coal Mine

The coal model adds reclamation cost to the basic strip-mining cost estimates. These costs were calculated from R. Evans and J. Bitter, *Coal Surface Mining Reclamation Costs* (U.S. Bureau of Mines, 1975) and *Coal Age* (July 1977). The basic calculation converted cost per ton by region estimated in these studies to cost per cubic yard of overburden. These costs were escalated to 1977, and the cost per cubic yard was then added directly to equation (2.10b) of chapter 2. The results by region are given in table C.1.

The studies examined mines where reclamation practices were equivalent to practices required under current law. Subsequent study by ICF Inc. in "Energy and Economic Impact of H.R. 13950," February 1, 1977, indicated small incremental costs because of the law.

Table C.1 Cost of reclamation ($)

Region	1	2	3	4	5	6
Cost per cubic yard	0.28	0.30	0.23	0.003	0.03	0.003

Table D.1 Transportation costs, including rents, in 1975 (1977$/ton)

Supply	Demand			
	New England (Boston) 1	Mid-Atlantic (New York) 2	East north central (Chicago) 3	West north central (Kansas City) 4
1 Northern Appalachia, Wheeling, Virginia	7.91	5.99	5.28	8.36
2 Southern Appalachia, Lynch, Kentucky	10.16	8.57	5.71	7.78
3 Midwest Harrisburg, Illinois	12.11	10.58	4.47	4.30
4 Montana-Wyoming (Powder River Basin), Billings, Montana	25.03	23.76	14.41	11.97
5 Colorado-Utah (Southwest), Price, Utah	24.57	23.31	13.25	10.90
6 Arizona-New Mexico, Gallup, New Mexico	27.89	26.38	16.90	9.14

Note: N indicates shipments eliminated for computational ease.

South Atlantic (Charlotte, North Carolina) 5	East south central (Chattanooga) 6	West south central (Houston) 7	Mountain (Grand Junction) 8	Pacific (San Francisco) 9
7.65	6.27	11.61	N	N
4.67	3.31	11.50	N	N
7.58	4.31	8.08	N	N
24.24	20.08	19.58	9.20	15.93
23.67	19.52	14.31	2.32	10.91
21.32	17.17	9.55	6.14	8.92

Notes

Notes to Chapter 1

1. The rising cost of nuclear power plants led to a decline in the relative attractiveness of nuclear power up to about 1973. The decline was not dramatic, but indicated that coal still had a role to play.

2. The Power Plant and Industrial Fuel Utilization Act of 1978 was signed into law in November 1978. It specifies that new electric power plants cannot be constructed to burn gas or oil as their primary fuels. New major fuel-burning installations are also prohibited from using oil and gas. Finally, existing power plants are prohibited from using gas after January 1, 1990. The law allows the Department of Energy to establish classes of exemptions.

3. In 1978 utilities accounted for 72 percent of total coal consumption in the United States.

4. The average lost workday for nonfatal injuries per 200,000 hours worked is 10.4 for all underground coal mining. That rate is exceeded only by the lumber and wood products industry. However, the severity of underground mining accidents, as measured by days lost, is twice as great for underground coal mining than for lumber and wood products. (Source: U.S. Labor Department Statistical Summary)

5. There have been several studies of the effect of the law on productivity.

6. There is currently an effort under way in Congress to repeal this act, but it is not likely to succeed.

7. This was the result of a major study of the costs of compliance with the act done by ICF, Inc., a consulting firm, for the Council on Environmental Quality and the EPA. See [5].

8. A challenge to the tax as violating the "undue burden on interstate commerce" clause was filed in the District Court of the First Judicial District of the State of Montana in the County of Lewis and Clark by a group of coal companies and electric utilities in 1978.

9. The original quote on transport rate was $7.90 per ton. This was to be escalated according to an agreed-upon escalation formula which would have taken it to $8.68 per ton in 1974. At that point Burlington Northern quoted a rate of $11.09 per ton. It is now at close to $20 per ton.

10. The author has developed a forecasting tool based on the supply and distribution model in conjunction with Data Resources, Inc. See [3].

Notes to Chapter 2

1. The supply model methodology has been discussed in [24] and [26]. The transport model is described in [25]. The demand model is an adaptation of the Regional Electricity Model documented in Baugham et al. [2].

2. The coal model used by the Department of Energy and developed by ICF, Inc. has a great deal of detail, but running a base case takes two CPU hours from a cold start. That is fifteen times the cost of the models described here. I am indebted to V. J. Chandru of MIT for this information.

3. The supply model excludes the following coal-producing states: Texas, North Dakota, Arkansas, Georgia, Iowa, Kansas, Missouri, Oklahoma, Washington, and Alaska. Only the first two states produce sizable amounts of coal. But the coal is of such little heating value that transport cost per Btu becomes prohibitive. Thus coal in those states is purely of local value. The production is set in these two states as a fixed fraction of local demand. The other states (excluding Texas and North Dakota) account collectively for less than 3 percent of national output. Furthermore this production is also of primarily local value.

4. The results by state are very sensitive to initial tax assumptions. In several regions substitutability between states is very large. Small differentials in initial tax levels can cause large shifts in state breakdowns. This is so because the long-run equilibrium nature of the model assumes mines were opened knowing these differentials.

5. The weights were calculated using the demand model. In that model demand is calculated by states. The state data are used to calculate what percentage of demand in the census regions is accounted for by the smaller regions.

6. The cumulative cost curve first appears in Hotelling's theoretical analysis of exhaustible resources [7]. Hotelling considered the problem of the behavior of price over time when cumulative output led to cost increases.

7. This assumption is supported by engineering analyses. Comparing descriptions of continuous mining operations in the 1950s with estimates of the Bureau of Mines in 1970, and with estimates in the Mining Engineering Handbook of 1973, indicates little variation in composition of a section. More recent engineering analyses also suggest little capital-labor substitutability. See the analysis by the NUS Corporation [11].

8. This was confirmed by discussion with a large equipment manufacturer. A more complete analysis could treat other types of overburden removal equipment, but the results would be only slightly altered.

9. See the American Institute of Mining Engineers handbook [1].

10. See Stefanko et al. [15].

11. The data are from mining inspection reports of the Mine Health and Safety Administration, formerly MESA. Data on depth of seam were not available for this sample. A smaller sample where data on depth were available was tested. Results indicated depth was not a significant determinant of productivity. See Zimmerman [24].

12. Log ϵ_i is normally distributed by assumption. Therefore the probability of log ϵ_i is given by:

$f(\log \epsilon_i) = 0$ if $\log \epsilon_i < \log T - \log A - \gamma \log Th - \beta \log S - \alpha \log OP$

$$f(\log \epsilon_i) = \frac{\phi\ (\log \epsilon)}{1 - \displaystyle\int_{-\infty}^{\log A - \gamma \log Th - \beta \log S - \alpha \log OP} \phi\ (\log \epsilon)}$$

if $\log \epsilon_i \geq \log T - \log A - \gamma \log Th - \beta \log S - \alpha \log OP$.

Jerry Hausman[6] provided a program for this estimation.

13. The cost of capital was calculated in the following way:

$$I = \int_0^T (c - \text{taxes})e^{-rt}\, dt,$$

where
I = initial investment,
c = annual return necessary to realize an after-tax return of r percent,
T = life of capital good.

Let
u = corporate income tax,
v = depletion allowance in percent of gross profit
$\quad = v(c - \text{depreciation})$.

Then

$\text{Taxes} = u(c - \text{depreciation} - \text{depletion})$,

$$I = \int_0^T [c - u(c - \text{depreciation} - \text{depletion})]e^{-rt}\, dt$$

$$= \int_0^T ce^{-rt}\, dt - \int_0^T uce^{-rt}\, dt + \int_0^T u(\text{depreciation})e^{-rt}\, dt$$

$$+ \int_0^T u(\text{depletion})e^{-rt}\, dt$$

$$= cF + ucF + u\int_0^T (\text{depreciation})e^{-rt}\, dt + u\int_0^T r(c - \text{depreciation})e^{-rt}\, dt,$$

where

$$F = \int_0^T e^{-rt}\, dt.$$

Then
$I = cF - ucF + uvcF + (u - uv)(\text{present value of depreciation}) \qquad (1)$
The last term on the right-hand side is

$$(u - vu)I\frac{2}{rt}[1 - \frac{1}{rt}(1 - e^{-rt})].\tag{2}$$

See Hall and Jorgensen [5].

Equation (2) is substituted in (1), and the resulting equation is solved for C/I:

$$\frac{C}{I} = \text{cost of capital} = \frac{[1 - (u - uv)(1 - (F/t)2/rt)]}{F(1 - u + uv)}$$

14. UMW fee is taken as \$1.385 per ton beginning 1978. Earlier years it is \$.82 per ton.

15. The estimated equation was used to solve for the optimum number of openings. However, given the results, the solution was to take the minimum number of openings possible. This we assume is two, one for intake of miners and supplies and one for removing coal. The cost of these shafts is added to total capital cost. The cost was taken as \$6,217,500, as reported by Williams [23] in 1974, and adjusted to 1977 prices, using BLS indices.

16. The constant 0.89 arises in the following manner:

$$\text{Total cubic yards removed} = \frac{\text{feet of overburden} \times \text{acres mined} \times 43,560}{27},$$

where
    $43,560$ = square foot per acre,
        27 = cubic foot per cubic yard,
acres minded = $Q/(Th \times 1,8000)$,

where
    Q = annual output,
    Th = thickness of seam in feet,
    $1,800$ = tons per acre foot.

Thus

$$\text{Total cubic yards} = \frac{R \times Q \times 1,800 \times 43}{27} = 0.89RQ.$$

17. See Bureau of Mines [17]. The basic data were escalated to 1977 prices, using BLS construction machine index. Contour mines were excluded. Furthermore the mines are based on 1969 practice which includes modest levels of reclamation. Reclamation costs are added separately, and the estimation is described in appendix C.

18. The Montana data were not recorded for higher overburden ratios. The cutoff ratio decreased as depth increased. This truncated the upper tail of the distribution, accounting for the curvature in the upper tail. Thus the approximation is better than the data show. The graph presents only the Powder River Basin data.

19. For a description of their technique see appendix G of [21].

20. See Zimmerman [25] for a documentation of regulatory procedure and a fuller development of the model. The variable J was not part of the model reported in [25].

21. The sum of $\alpha_5 + \alpha_6$ is 0.055. The average Btu content of coal is 24 million Btu in the east. *PALT-FOBCOAL* is in cents per million Btu. The tariff, T, is measured as $/ton. A $.01 change in *PALT-FOBCOAL* therefore corresponds to a change of $.24 per ton of coal equivalent. The coefficient α_5 therefore must be between 0 and 0.24.

22. There is no clear separation between metallurgical coal and steam coal. In eastern markets coal that was normally used to produce coke has entered the steam market because of the need for low-sulfur coal to satisfy environmental pollution regulations.

23. George Rozanski designed and programmed this model change.

24. For an attempt to make the model fully dynamic, see [22]. This work attempts to integrate a much smaller version of the coal supply transport and distribution model with an optimizing utility model, using Bender's decomposition algorithm. This approach needs more work before it is available for policy analysis.

25. The steps are chosen so as to produce no more than a maximum divergence of 1 percent between actual cost as measured by the cost function and the cost on the step function.

26. Retrofit capital cost is taken as 60 percent higher than new construction. This figure is the midpoint of the range of estimates reported by Gordon [4], appendix C.

27. We assume 90 percent of coal shipments are not renegotiated each year. There are no systematic data available on exactly how much coal is shipped in each year based on previous commitments. An item in *Coal Week* on April 13, 1976, indicates that in any year TVA negotiates no more than 10 percent of its burn.

28. Restricting future shipments in the transport model leads to anomalous results. The model makes decisions on current costs. Fixing future shipments based on current costs leads to substantial production of high-cost reserves, when less costly reserves are available. Preliminary testing of the dynamic version [22] indicates that when future run-up in cost is taken into account, the results are closer to those we report than to the results with commitments binding in the future.

29. We have assumed a single type of scrubber. As chapter 1 indicates, 70 percent scrubbing is possible on low-sulfur coal. Our results must be viewed as a limiting case. The net difference between our assumption and actual regulations will be small. On this last point see Data Resources, Inc. [3] for a study of alternatives using the same coal supply model described here.

30. The bias involved in this treatment of sulfur constraint can be calculated as follows: for the first 20 years (1975 to 1995) incremental demand is

$ID = D(t) + \delta D(0)[0.05t - 1]$,

where $\delta = 0.9$. The NSPS demand will equal demand in new capacity plus replacement of obsolete coal capacity. Assume that $D(t) - D(t - 1)$ represents new capacity and refinements are 5 percent per year. Then

$$\text{NSPS demand}_t = D(t) - D(t-1) + 0.05 \cdot D(0) \cdot t,$$
$$\text{Bias} = (\text{NSPS demand} - ID)$$
$$= D(0)[-0.9 + 0.005 \cdot t].$$

In 1975 nonmetallurgical demand was 486.4 million tons. This yields the following year-by-year calculation of the bias:

	D(t)	Bias	Percent
1975	486		
1980	544	− 36.5	− 6.7
1985	705	− 24.3	− 3.4
1990	767	− 12.2	− 1.6

Notes to Chapter 3

1. We are being optimistic and assume that safety related shutdowns become rarer and utilization rates rise. We test sensitivity of our results to this assumption in chapter 5. The analysis of that chapter indicates that a 65 percent operating rate would not substantially affect our conclusions. Furthermore we set a maximum availability rate of 73 percent. Actual operating rates of utilities have been below their actual availability factors, so we do not have too great an optimism here.

2. We use 1973 data because breakdowns of production by sulfur content are available. Further 1973 was a year in which structural changes in the coal industry began to be felt.

3. The most important of these suits was a challenge brought by the Sierra Club to prevent leasing of western coal lands before an environmental impact statement for the entire program was prepared. The suit was settled in 1977.

4. In 1975, for example, only 52 million tons of coal were consumed west of the Mississippi.

5. This is true of mine-mouth prices as well. The cheapest coal is mined first regardless of sulfur content.

6. This is the case for demand regions 7 and 8.

7. This assumes 17 million Btu per ton for coal in the mountain states and 24 million Btu on the eastern seaboard.

8. For discussion and justification see chapter 2 and note 7 of chapter 2.

9. They were entered in smoothed annual increases to cumulate to a 34.8 increase by 1980.

10. The average annual real increase in wages since 1970 has been 2.8 percent. We have assumed a slowing down to keep a conservative bias. The increase can be regarded as the net impact of labor productivity change and wages. Given the historical record on the net effect, we are presenting a very optimistic scenario.

11. This indicates the UMW has more to fear from sulfur pollution regulations than from nonunion mines in the west.

12. The average cost of coal burned by electric utilities in the United States rose by 62 percent between 1973 and 1974. See *Statistical Yearbook* [2].

13. The national average price of electricity rose by 19 percent between 1973 and 1974. See *Statistical Yearbook* [2].

14. The Department of Energy is now forecasting a 4.2 percent average annual rate of growth of electricity demand between 1977 and 1995. EPRI forecasts 5.3 percent increase in annual growth rate through 2000. See *Annual Report to Congress* [6].

15. These model runs all assume oil prices that are low, considering the most recent developments. Thus actual rates of growth will be below those shown here. Nevertheless, the main point is the relative insensitivity to coal supply developments. For a discussion of this point in the context of model integration methodology, see Baumann et al. [1].

16. More recent oil prices would simply reinforce this fact.

17. The range in current forecasts of nuclear power for the year 2000 is 200 to 300 GW. Our figures are high since they are based solely on observed costs of construction and operation. Regulatory uncertainty and public opposition are not represented in these estimates. For a discussion of this issue see chapter 5 and references cited therein. For the range of nuclear forecasts, see [6].

18. Five years ago (1975) government forecasts of nuclear capacity for 2000 were in the range of 425 to 565 GW. By 1978 the forecasted range was 255 to 395 GW. Current government forecasts are 235 to 300 GW, which reflect regulatory difficulties. Current market surveys by reactor vendors see a 2000 capacity of 250 to 325 GW. See *Annual Report to Congress* [6].

19. For a discussion of the effects of regulatory hassles on nuclear power, see Zimmerman and Ellis [9].

20. The forecasting algorithm of the Baughman-Joskow model means that rising prices for low-sulfur coal are taken as evidence that prices will continue to rise. Thus the run-up in low-sulfur coal prices leads to expectations of a still higher differential.

21. Given that the oil prices used here are low considering recent developments, this trend will accelerate.

22. The range of forecasts as of 1979 for 1990 industrial coal consumption (including metallurgical use) is 214 to 300 million tons. Our estimate, including metallurgical coal, is 237 million tons.

23. Fluidized bed combustors present problems of their own. They use large quantities of limestone or dolomite to remove sulfur from the fuel, and this material must be disposed of.

24. For a development of this formula see Herfindahl and Kneese [4], chapter 4, or Modiano [5].

25. White [7] has utilized Bender's decomposition algorithm to do this on a much reduced version of this coal model.

26. This result is confirmed in the White's study [7]. High rents accrue to those segments that supply the exogenous demands. A lower interest rate would lead to higher costs. The cost of capital to the coal industry has in fact been higher than 10 percent. See Chris Alt, "Cost of Capital to the U.S. Coal Industry" (MIT Energy Laboratory, mimeo, 1978).

Notes to Chapter 4

1. Real capital costs are increasing at 2 percent per year; real wages at 3 percent per year; and real transport costs at 3 percent per year.

2. This is higher than the impact estimated in chapter 2. The reason is that here transport costs are rising and interact with sulfur pollution regulations.

3. At the time of this writing the EPA is considering what regulations to impose on large industrial users of coal. In the absence of regulations forcing scrubbing, industrial users would use low-sulfur coal. For information on this last point, see Dyck et al. [4].

4. See appendix table B.5 for totals for oil metallurgical and export coal.

5. The Fuel Utilization Act of 1978 actually does prohibit oil and gas use in new boilers. However, there are two exemptions available. Current policy is to exempt users where environmental laws would be violated or the cost of compliance is 30 percent greater than the cost of oil or gas. The 30 percent is an interim figure. See U.S., Department of Energy [9].

6. In chapter 5 we examine what happens if nuclear power is at a higher cost or not available.

7. Sulfur dioxide is feared to have deleterious effects on animal and plant life. Current concern is great over the effects of acid rain. The SO_2 mixing with moisture in the atmosphere forms acid rain that can pollute lakes. For a discussion of the health problems associated with SO_2, see Ramsay [6].

8. We are comparing costs of current standards to no standards. However, the bulk of costs associated with sulfur reduction to 1.2 lb are in the 2.2 to 1.2 lb range, as we have seen.

9. *Statistical Abstract of the United States* [8], series IIB, p. 16.

10. These are costs to consumers, not measures of social cost. Electricity is priced according to average cost formulas. We take these as given in the model and look

only at impact on particular groups. Transmission of SO_2 over wide areas makes it impossible to compare costs with incidence of pollution.

11. See appendix table B.1 for a list of current taxes expressed on a percentage basis.

12. See Alt and Zimmerman [1].

13. Current oil imports are roughly 7 million barrels per day at an average price of about $30 per barrel. Total coal production in 1978 was 654 million tons at a total value of $14.6 billion.

14. Furthermore no allowance is made for the inefficiency involved in underpricing electricity. Our calculation takes the present regulatory system as given.

15. President Nixon started with Project Independence in 1974.

16. This conclusion is reinforced when one considers that the assumed oil prices are low. Higher oil prices most likely mean oil demand is less sensitive to coal prices.

17. Externalities mean environmental costs not now internalized by the Surface Mine Reclamation Act.

18. Current plans call for leasing to resume in 1981. However, there are significant policy debates about how much to lease and under what terms leasing will resume.

19. It will be particularly costly if EPA finds these wastes are hazardous and disposal thus becomes subject to rigorous standards.

20. Evidence suggests that the net difference between full BACT and the 70 percent option is small. This is because the incremental costs of 90 percent removal are not that great. Only a relatively small benefit is provided for western coal in this way. See Data Resources, Inc. [3].

21. The total cost estimate is similar to an EPA estimate of the costs of BACT in 1995 [5]. The lowering of price in some regions seems paradoxical. However, two effects are at work. First, BACT lowers the cost of low-sulfur coal. In regions where low-sulfur coal is consumed without BACT, costs are thus lowered. Second, the utility model bases decisions on the expected present value of costs of alternative generation types over the entire life of the plant. Thus, while the present value of costs goes up under BACT, in any given year the regulated average price of electricity can be lower. The author is indebted to Paul Joskow for pointing out the latter fact. For a similar effect when the utility model is used alone, see Baughman et al. [2], p. 176.

22. Because of the lower heating value of western coal, in order to meet a 1.0 lb SO_2 standard, Powder River Basin coal would have to have a sulfur content of 0.425 percent by weight. Midwestern coal of 4 percent sulfur would meet a 1 lb SO_2 standard with 90 percent scrubbing. To check this contention, we ran a scenario without BACT and a 1 lb standard. The results were very close to the scenario presented in the text.

23. The checkerboard pattern results from federal land grants to railroads.

24. For a detailed analysis of leasing strategies and procedures, see Tyner et al. [7].

25. Recall from chapter 3 that depletion in strip reserves is much more rapid than deep reserve depletion in the midwest.

26. This is still western coal, and the use of the term "high-sulfur" here indicates only that the coal is used in plants with scrubbers. The average sulfur content is still far below the level of high-sulfur eastern coal.

27. The Louisville and Nashville railroad succeeded in getting approval for a 28 percent increase in coal rates over the last twelve months.

28. See Alt and Zimmerman [1] for a more detailed analysis of taxes, which assumes higher oil prices, higher nuclear costs, and lower growth rates. The consequently lower demand growth for electricity led to an optimal tax of approximately 62.5 percent. Thus in any case there is substantial market power.

29. We subtract the real cost increase of the base case from the 30 percent increase in calculating the railroad rents.

30. A national severance tax or joint action on raising rail rates are examples of the latter possibility.

Notes to Chapter 5

1. The 65 percent capacity factor is a current forecast.

2. For a discussion of these issues and an attempt at measuring these implied costs, see Zimmerman and Ellis [5].

3. A probability model estimated by the author estimates that the probability of building a nuclear plant in the midwest, assuming the utility needs a baseload plant, is zero. See Zimmerman and Ellis [5].

4. The effect of a nuclear moratorium is to raise total rents but not change the tax that maximizes revenue. See Zimmerman and Alt [6].

5. Current knowledge of CO_2 absorption in the biosphere is inadequate to predict when we will face climate changes. See National Research Council [2].

6. Current forecasts for long-term rates of growth in electricity demand range from a Data Resources, Inc., forecast of 2.8 percent, Data Resources, Inc. [1], to a DOE forecast of 4.2 percent, Department of Energy [3], and a forecast by EPRI of 4 percent (their *low* case) cited in *Annual Report to Congress 1978* [3], p. 275.

7. Again, however, the net difference between partial scrubbing and full scrubbing appears to be small.

Notes to Chapter 6

1. In the moratorium case total electric utility consumption is 45.174×10^{15} Btu in 2000. Total coal consumption, including metallurgical and export, is 55.14×10^{15} Btu.

2. This will be the first installment in a total aid package that ultimately could reach $80 billion.

3. See Department of Energy [1], chapter 13. It should be noted that many observers feel that this high a level of synfuel production will not be reached by 1990 or 2000.

4. Synfuel projects, due to political considerations, will be allocated among regions. However, large-scale development would take place where the economics are most favorable, which clearly will be in the west, as transport of gas or oil will be relatively cheap. Substantial exports of low-sulfur steam coal could only come from the west for reasons discussed in chapter 3. The upper range of WOCOAL forecasts represents the engineer's view of what could be done.

Notes to Appendix A

1. The distribution and the numerical approximations were worked out by Michael Baumann.

2. There is some dispute as to whether a maximum overburden ration criterion was also applied to the strippable reserve base. For evidence that it was, see Zimmerman [1].

Index

Acid rain, 160
Air pollution standards, 12, 101–105
cost of, without nuclear power, 164–165
See also Sulfur emission
American Association of Railroads Cost Index, 43
Appalachian coal, 2, 3, 9, 11–12, 26, 61, 62, 70, 93, 103, 111, 123, 124, 166, 170–171
decline of, 67–69
Appalachia, strippable reserves in, 179–180
Arizona–New Mexico coal, 104
BACT, 132, 133, 137
leasing moratorium and, 139

BACT, (Best Available Control Technology,) 12, 63, 64, 88, 130–137, 140, 144, 147
without nuclear power, 164, 165, 168
Baughman, Martin, 46
Baughman-Joskow-Kamat model, 46, 48, 52, 152, 153
Bituminous Coal Act, 1937, 19
Black lung disease, 10
Burlington Northern Railway, 15

California, 14
Carter administration, 1, 107
Clean Air Act, 1967, 1, 59,
1977 amendments, 12, 13
1978 amendments, 130
Coal
consumption of, 5
cost of, to industrial consumer, 123
future of, 170–178
industrial demand for, 89–91
industrial use of, 106, 107
Coal Mine Health and Safety Act, 1969, 1, 10, 24
Coal policy
interaction with nuclear policy, 161–164
oil imports and, 173–175
Coal refuse, disposal of, 9
Colorado, 3
strip mining in, 41–42
Continuous mining, 25
Cost functions, 27–28
deep-mining, 28–30
strip-mining, 31
Costs
labor, 62
user, 91–93

Deep mining, 3, 5, 9, 10, 72
cost functions, 28–30
expenditures, 30–31
technology, 25
Demand
factor price changes and, 82–91
regional aggregation, 21–22
Demand constraint, 53–54
Demand model, 22–24, 45–50
link with supply model, 50–56
regulatory section, 49–50
Demand shifts, effects of, 64–72
Department of Energy, 13
Depletion, 24, 64–72
Distribution, cost of mining and, 40–41
Dragline, 31–32, 33*f*

Electricity
cost of, to consumer, 119–123
prices, 88–89, 144
Electricity demand, 46, 82–83
effect of nuclear moratorium on, 161
Electricity production, 1970s, 1
Electric utilities, 5

Electric utility model, nuclear power generation and, 152–155
Electric utility sector, demand for coal in, 46–49
Endogenous variables, demand model, 56
Energy policy, U.S., 100
Environmental regulation, 1–2
Environmental trade-offs, 100–151
EPA, (Environmental Protection Agency), 11, 13
Estimation procedure, 26–28
Exogeneous variables, demand model, 56

Federal coal lands, leasing policy, 137–145
Federal Energy Regulatory Administration, 153
Fuel choice, alternative sulfur standards and, 106–116
Fuel Utilization Act, 1978, 171

Health and safety, deep mining and, 10
Health and Safety Act. See Coal Mine Health and Safety Act
High-sulfur coal, 104, 105

ICC (Interstate Commerce Commission), 14, 42
Illinois, 13, 39, 39f
Illinois Basin, 3, 80, 71, 72
Internal Revenue Service, national coal tax, 14

Joskow, Paul, 46

Kamat, Dilip, 46

Labor productivity, trends in, 62
Leasing moratorium, western coal under, 166
Leasing policy, federal coal lands, 137–145
Low-sulfur coal, 10, 11, 14, 102, 103, 104

Marginal cost, long-run, 34–36
Metallurgical coal, 50
Midwestern coal, 65–67, 71
decline in output, 70
prices, user cost and, 93
Mining, continuous, 25
Models, 17–58
Montana, 3, 145
severance taxes, 14
Montana-Wyoming coal, 64–65, 70, 75, 76–79, 101, 110–111
BACT, 132, 137
leasing moratorium and, 139
prices, user cost and, 93
Municipal Electric Utility, San Antonio, 15

National parks, nondegration standards, 12
New England, 68, 102, 103, 107, 108, 111, 122
New York State, 14
Nondegration standards, 12,
NSPS (New Source Performance Standards), 12–13, 54, 55
Nuclear moritorium, 155–168
coal industry under, 152–169

effect on coal prices, 156
effect on electricity prices, 157–160
Nuclear power, 1, 2, 13–14, 60
coal industry without, 152–169
coal versus, 83–88, 109–110
generation, predicting, 152–155
moratorium, 155–168
Nuclear Regulatory Administration, 152–153
Nuclear Regulatory Commission, 14

Ohio, 13
Oil imports
coal policy and, 173–175
sulfur regulations and, 128–129
OPEC oil embargo, 1973, 1
Overburden ratio, 36, 39–40
Overburden removal, 31–32, 33f. See also Overburden ratio

Pennsylvania, 12
Pike County, Kentucky, 37–38, 38f
Pneumoconiosis, 10
Policy changes, effect of, 41–42
Policy variables, demand model, 56
Pollution
at consumption point, 10
at mining site, 9–10
See also Sulfur
Powder River Basin, Montana, 39, 69, 71, 72, 156
BACT, 133
taxes on coal, 145
Prices, coal

changes, 69
effect of nuclear moratorium on, 156
Prices, electric,
88–89, 144
effect of nuclear moratorium on, 157–160
Production, regional
effects of alternative
sulfur emission levels
on, 110–116
taxes on, 145
trends, 5, 9

Radioactive waste, disposal of, 14
Railroad rates, 5,
42–45
eastern, 43–44
increases, 14–15, 145
midwestern, 43–44
western, 44–45
Reclamation, costs,
11–12, 189
REM (Regional Electricity Model), 46.
See also Baughman-
Joskow-Kamat model
Rents, 14–15, 91–93,
145–147
Respiratory ailments,
coal emissions and, 10

San Antonio case, railroad rates, 15
Scrubbing, 10–11,
13, 22, 88, 107–108,
131
capacity, 108
Seam thickness, 36–39
Severance taxes, 14, 145
Sierra Club, 138
SIP (State Implementation Plans), 13
Southern Company,
Virginia Electric
Power Company, 15
Stack gas scrubbing,
10–11, 22, 88, 131
Steel making, coal for,
50

Strip-Mine Reclamation
Act, 1977, 11
Strip mining, 3, 5, 9–10,
11, 41–42, 72–73
cost functions, 31–34
expansion, 147
expenditures, 34
government regulation
of, 1
technology, 25–26
Sulfur
aggregation of, 22–24
constraints on, 54–56,
70
price effect of, 71–72,
85, 89
Sulfur alternative standards, effects of on
fuel choice, 106–116
Sulfur emission standards
and coal prices, 101–
105
NSPS and, 12–13
Sulfur pollution, 1, 3, 5,
9, 10–11
effect on tax revenues,
125
indirect costs, 127–128
net direct costs, 127
oil imports and, 128–
129
reduction of, 10, 116–
128
self-regulation of, 69–
71, 88–89, 147
Supply
developments, 79–82
regional aggregation
of, 19–21
Supply model, 24–25
link with demand
model, 50–56

Taxes, severance, 14,
145
Tax revenues
Montana-Wyoming
coal, 146f
sulfur regulations
and, 125

Technology
deep-mining, 255
strip-mining, 25–26
Three Mile Island, 2,
14, 84, 152
Transport model, 42–43
Transmission and distribution, 50
Transport rates, 42–45,
76–79, 85, 145, 147.
See also Railroads

Utah, 3
Utah-Colorado coal,
104
BACT, 132, 137
leasing moratorium on
and, 139
United Mine Workers,
30–31, 81, 82
1979 contract, 74
United States Geological
Survey, 36, 40
User costs, 91–93

Wages, 74–75
Water contamination, 9
Western coal, 64–65.
See also Montana-
Wyoming; Utah-
Colorado
Western mining
nuclear moratorium
and, 166
reducing, 129, 130
Wisconsin, 14
Wyoming, 3
severance taxes, 14,
145

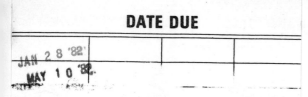